TEACHERS' MANUAL OF BIOLOGY

Maurice A. Bigelow

Professor of Biology in Teachers College,
Columbia University.

MAVEN BOOKS

Chennai   Trichy   Tirunelveli   New Delhi

MAVEN BOOKS

An Imprint of MJP Publishers

ISBN 978-93-87867-06-2   **MAVEN BOOKS**

All rights reserved   No. 44, Nallathambi Street,
Printed and bound in India   Triplicane, Chennai 600 005

MJP   572   © Publishers, 2019

Publisher: **C. Janarthanan**

PUBLISHER'S NOTE

The legacy of a country is in its varied cultural heritage, historical literature, developments in the field of economy and science. The top nations in the world are competing in the field of science, economy and literature. This vast legacy has to be conserved and documented so that it can be bestowed to the future generation. The knowledge of this legacy is slowly getting perished in the present generation due to lack of documentation.

Keeping this in mind, the concern with retrospective acquiring of rare books has been accented recently by the burgeoning reprint industry. Maven Books is gratified to retrieve the rare collections with a view to bring back those books that were landmarks in their time.

In this effort, a series of rare books would be republished under the banner, "Maven Books". The books in the reprint series have been carefully selected for their contemporary usefulness as well as their historical importance within the intellectual. We reconstruct the book with slight enhancements made for better presentation, without affecting the contents of the original edition.

Most of the works selected for republishing covers a huge range of subjects, from history to anthropology. We believe this reprint edition will be a service to the numerous researchers and practitioners active in this fascinating field. We allow readers to experience the wonder of peering into a scholarly work of the highest order and seminal significance.

MAVEN BOOKS

INTRODUCTORY NOTE

Tʜɪs *Manual* for teachers has been prepared in order to give more fully than was possible in the *Applied Biology* for students, the author's suggestions regarding the use of that book, especially in many cases where there is a radical departure from the usual teaching of elementary biological courses. Moreover, this *Manual* aims to be of help to biology teachers whose special preparation has been limited. Much of the weakness of the present-day high-school biology is due to the fact that the majority of teachers have not critically studied the integral parts of the courses that they are trying to teach. Textbooks of language and mathematics are carefully analyzed for the teacher, either in appendices or in separate " keys " ; and for the teaching of biology there is need of some such guide to give specific advice concerning methods, time, materials, relative value, etc., in teaching difficult topics.

This *Manual* does not, in any way, take the place of the " Teaching of Biology in the Secondary School," by F. E. Lloyd and M. A. Bigelow, published by Longmans, Green, and Company in 1904. That work deals entirely with general principles, while this is concerned with the detailed working out of lessons in a particular type of elementary course in biology. Wherever possible in this *Manual*, specific references are given to the appropriate discussions in the " Teaching of Biology," thus saving the time of teachers who wish to refer to that book.

The chapters and the numbered sections of this *Manual* correspond with those of the *Applied Biology*. An appendix in the *Introduction to Biology*, which is in preparation, will contain a table for quick reference from any numbered section of the *Introduction* to sections of the *Applied Biology* and of this *Manual* that deal more fully with the same topics.

A list of books for teachers' reference will be found in an appendix to this *Manual*. Authors' names are arranged in alphabetical order, and the date of publication (copyright) of each book is added to avoid confusing past or future revised editions. References to this list are by names and dates; for example, Lloyd, 1904, 49, refers to page 49 of *The Teaching of Biology*, by Lloyd and Bigelow; while Bigelow, 1904, 261, refers to page 261 of the second part of the same book, which has entirely independent parts for botany and zoölogy.

The index to this *Manual* will make it useful to teachers who are not working with the *Applied Biology*, but who want practical points concerning the teaching of topics in other textbooks. If the *Manual* is used for this purpose, the corresponding sections of the *Applied Biology* should be examined at the same time, for the *Manual* is largely a teachers' supplement to the accompanying textbook.

The author will gladly be of service to any teachers who use this *Manual* or the *Applied Biology*. From time to time, until a reprint of this *Manual* allows supplementation, the author proposes to issue mimeographed sheets containing new suggestions. These will be mailed, without charge, to teachers who file self-addressed and stamped envelopes with the author. Proper credit will be given teachers who send helpful criticisms or suggestions.

The authors of the *Applied Biology* are already greatly indebted to the following biologists who have given suggestions that are incorporated in this *Manual:* Professor W.

F. Ganong, of Smith College; Professor W. S. Hall, of Northwestern University Medical School; Professor R. E Schuh, of Howard University; Dr. C. Stuart Gager, Director of Brooklyn Botanical Garden; Miss Anna M. Clark, of New York Training School for Teachers; Professor T. D. A. Cockerell, of Colorado; Professor John Dearness, of London, Ontario.

M. A. B.

TEACHERS COLLEGE,
  COLUMBIA UNIVERSITY,
    May, 1912.

CONTENTS

CHAPTER I[1]

BIOLOGY AS A SCIENCE

1. *Simple classifying.* — It is well worth while to ask young students to write in parallel columns the names of various living and lifeless things. Such a problem will strike many teachers as absurdly simple. It so impressed the writer before it was proved by many trials with high-school classes that most students can profit by critical thinking on even so simple a classification as that into living and lifeless. The problem is well worth five minutes for serious consideration by the students, preliminary to summarizing by the teacher. The best results have come when the teacher placed on the blackboard the names of things suggested by students, arranging them in columns for lifeless and living. As a side issue, it is worth while to help the young students do some simple classifying; for example, many lists written by them read: "birds, robin, chicken," or "oak, maple, trees," or even "plants, sunflower, corn." In all such cases the difference between specific names and collective names of groups of natural objects should be made clear by grouping the specific names with brackets on the blackboard. The writer has on several occasions seen many minutes profitably spent on such simple classifying, which the students' notes showed was needed. *F*inally, it will add interest to note some of the doubtful cases, such as dried seeds capable of germinating; but it would be a useless exercise

---

[1] The chapters and sections of this *Manual* are numbered to correspond with those of the *Applied Biology*.

to discuss at this stage whether such things are dead or alive.

3, 4. *Definitions of sciences.* — At the outset, students should have some idea of the field of any new study. Experience shows that in biology they cannot be left to form their own definitions. Especially should teachers guard against the common tendency, which appears even in colleges, to use zoölogy and biology as synonymous terms. Emphasis upon § 4 in the *Applied Biology* will avoid such a serious misunderstanding.

While various attempts have been made to limit the term "biology" to special usage, such as to bio-physiology, the science of life-activities, such limited use no longer need be regarded so far as general education is concerned. The word "biology" has now a widespread use as the general term including the fields of both botany and zoölogy, and biologists must invent new designations for special limited aspects of the general field of biology.

5. *Applied biology.* — This paragraph, in the *Applied Biology*, includes a plea and defense for the study of biology, and is an anticipation of the usual question: "Why should I study animals and plants?" Some preliminary glimpse of the outlook of biology in the direction of human life is sure to be helpful to many students who, without some idea of the goal, might see no practical meaning in a large part of the most useful subject-matter of biology. The oft-commended method of leaving the beginner to develop his own interest in biology or else succumb, while only the fittest survive, may possibly have its place in the training of special workers in the field of science; but it is essential that high-school biology should be conducted so as to awaken in the students the greatest possible interest. And this greatest interest is all-important because biology is so variously applicable to everyday human life, and not because the interest of students

means the professional success of the teacher. The wide-
awake teacher will take many opportunities, which are
opened but not developed in other parts of the text-
book, for impressing upon the students the fact that the
science of life is capable of extensive application in the
daily life of all citizens.

CHAPTER II

COMPOSITION AND CHANGES OF MATTER

6–11. *Essentials of chemistry and physics.* — Since the basis of correlation in the *Applied Biology* is physiological, it follows that some knowledge of the elements of chemistry and physics is necessary for the students. In many schools, and in some elementary text-books of biology and physiology, much attention is given to an extended series of chemico-physical experiments which are introduced in the first part of the course of biology to be studied several weeks before any reference is made to biological problems involving chemistry or physics. The objection to this practice is th t the students forget, and much chemico-physical teach g must be repeated when needed for application in ysiology, the result being wasteful duplication. For e mple, a preliminary series of chemico-physical lessons on foods must be repeated in essentials months afterwards when digestion is studied. A still stronger objection to such preliminary work is that a close correlation in time, as well as in subject-matter, makes both the chemico-physical facts and the biological applications more intelligible and more interesting. For example the best time for preliminary lessons on carbon dioxide is when a biological problem such as respiration calls for the appropriate chemical facts.

However, there are certain general facts, such as changes in matter, chemical elements, compounds, etc., which are not called for in biology in any special connection, and which, as a rule, must be treated as preliminary and without direct application. In the *Applied Biology* an attempt has been made to select such material for Chapter II, reserving as far as possible the chemico-physical topics that admit of close correlation with the biological problems of later lessons. The sole justification, then, for Chapter II is that it presents general ideas which are not connected with any particular lessons. The chapter would be unnecessary if all students could study elementary chemistry before biology.

The shorter course, *Introduction to Biology*, will include some directions for simple experiments, because it is assumed that the students who will use this book are in the early years of high school and probably have not had any chemistry and physics; hence, for them, the course in biology must be made an introduction to science in general. This is in harmony with the widespread idea that science in the first year of high schools is best when presented as advanced nature-study and as an introduction to general science, with human life as the center of interest. The experiments are omitted from the longer course (*Applied Biology*) because that is likely to be studied in high school later than the first year, and it is probable that the great majority of the students will already have had some introduction to the elements of the physical sciences, and hence the text of Chapter II will be more or less of a review of familiar facts from a new viewpoint. However, the experience of the authors of the *Applied Biology*, and that of other teachers who have been consulted, has been that very many high-school students take biology before chemistry or physics, and that most of those who have previously studied these sciences need a review of some of the fundamental ideas that are essential for the physiological phases of biology.

7, 8. *Changes of matter.* — The authors of the *Applied Biology* realize that the brief definitions of chemical and physical change in terms of "composition" of matter are open to some criticism by the specialists in physical chemistry; but for the purposes of practical life and of elementary teaching they are convenient and sufficiently accurate.

9, 10. *Elements and compounds.* — It is absolutely necessary for even the simplest physiological study that students have general ideas, at least, regarding elements and their part in the formation and disintegration of compounds. Even in college classes of biology there are frequently students who have not studied the simplest chemistry and who might be greatly helped even by such brief and superficial outlines of the essential facts as are given in §§ 9 and 10 of the *Applied Biology.*

CHAPTER III

CHARACTERISTICS OF LIVING THINGS

12. *Demonstration method for practical work.* — As stated in the *Applied Biology* in a footnote belonging to § 12, some of the practical work suggested in small type throughout that book is most satisfactory when demonstrated by the teacher. In most elementary courses of biology much time is often wasted and inaccurate impressions given the students by individual laboratory work on certain topics.

The decided advantages of demonstrations for some difficult problems are as follows: (1) Only one set of apparatus is required. (2) The teacher may be reasonably sure of working directly to the correct result, while a large percentage of students get incorrect results from their own experiments. (3) The teacher can make sure that all students in even a large class have seen the experiment correctly worked out, while this is impossible with numerous (20–40) students working individually. (4) Moreover, the teacher can lead the class as a unit through the successive steps of an experiment and can be certain of correct observations and conclusions. (5) The demonstration may be made an enormous saver of time, for often an energetic instructor can in ten minutes teach a practical lesson to the great majority of students so that they comprehend it at once far better than the exceptional few after a half hour's independent work.

The common objections to the demonstration method are as follows: (1) Students should get the technical training from performing the experiments themselves. One answer is that it is better for them to get scientific training from a limited number of problems selected because it is reasonably certain that the average pupil can solve them without waste of time. Another answer is that the average student does not need technical training in laboratory methods, but needs rather the information and point of view to be gained from a well-selected experiment. (2) Students are said not to "pay attention" to demonstrated experiments as well as to their own work. This is because science teachers have universally adopted a device for compelling attention in the individual work. That device is the laboratory notebook, which students respect because it is destined for the teacher's inspection; and it should be applied to demonstrations given by the teacher. Every student's record of a demonstration should give the following: (1) statement of problem to be solved; (2) description of teacher's method of attacking the problem, with labeled sketch of apparatus used; (3) results of the experiment; (4) conclusions with reference to the problem to be solved, or what the experiment has proved. If the experiment fails to give definite results, record under (3) any discoverable factors that may have caused the failure. With such required records for all demonstrations, there will be no difficulty in securing the closest attention.

After careful consideration of all phases of the problem of demonstrations, the author now believes that it is far better in elementary biology to put into demonstration form all difficult practical work which if taken up as individual problems means failure and waste of time for a majority of the average students in large classes.

13. *Substitute apparatus.* — A chemist's ignition-tube (a kind of test-tube of hard and thick glass) may be stoppered with clay through which is inserted a small glass tube or a pipe-stem. Or the teacher may use a short piece of gas-pipe, iron capped at one end, the other end stoppered with clay packed around an outlet glass tube or pipe stem.

17. *Activities of living things.* — Experience with this lesson has shown that it is desirable to put before the students at this stage of the course some very general ideas concerning the life of animals and plants in order to give some preliminary view of problems to be worked out more in detail. Without such a preliminary survey, the students must work blindly. The author is well aware that there are teachers in colleges and high schools who see no objection to working blindly for weeks or months, with the expectation of giving the facts some kind of an interpretation before the end of the course; but such a practice is entirely out of line with the whole trend of modern scientific education. The strict Agassiz method of leaving a student to work through a maze of intricacies of animal and plant structure, without the guidance of a definite suggestion as to what he is working for, may have made or discovered a few great biologists; but imitation of this method has done more than anything else to keep high-school and college biology from becoming, for the great majority of students, the popular subject that its content, which is so full of human interest, ought to make it.

Biology teachers can no longer afford to ignore the oft-repeated criticism by intelligent people who believe in biology for the purposes of liberal education, that so much work commonly assigned in biology seems to be without a definite goal and without any satisfactory meaning to students after the work has been accomplished. This is not an unfair criticism of much of the common laboratory and text-book work, especially along morphological lines. It is greatly to be desired that high-school biology, and perhaps other sciences, should be put into line with the educational principles that have been successful in the best nature-study teaching in the elementary schools; and this means developing the subject logically so that the student may see in advance the outlines of the path he is to travel, may understand the points of interest as he passes along, and at the end be able to look back along the way in a comprehensive retrospect. In other words, the student has a right to know before, during, and after each extensive exercise, just why he is studying that particular topic of biology. The writer must confess that this is a difficult idea to put into practice at all times, but it is an aim well worth careful consideration by the teacher of high-school biology.

The lessons in §§ 17–30 of the *Applied Biology* will undoubtedly call forth the criticism that abstract generalizations are beyond the students. The author replies that the avoidance of technical language makes it possible to state some of the greatest ideas of biology in very elementary form. It will probably be objected by some teachers that such general ideas regarding life-activities should come at the end of the course, after the students have acquired a large mass of facts on an inductive basis. This view is plausible in theory, but critical examination will reveal in §§ 17–30 little requirement for a basis of induction beyond what ordinary observation and the nature-study of elementary schools may be expected to give. Moreover, it is well to remember that the strict use of the scientific method is very limited in science teaching. At best we can do little more than show students a point at a time in one animal or plant type, and then tell them that, so far as is known in present-day science, the same thing is true of other — in some cases of all — animals and plants. (See discussion of scientific method in biology teaching by Bigelow, 1904, pp. 299–309.)

In order to get the greatest possible value that may come from study of such general considerations after the students have acquired more facts, the author recommends that §§ 17–30 in the *Applied Biology* should be reviewed just before the study of Chapter VI; but in order to give the students some outlook for Chapters IV and V, it is important that this lesson should be studied before proceeding with the more detailed work of those chapters.

18. *Automatic movement* is, of course, not absolutely distinctive, if we plunge deep into chemical and physiological analysis. Such movement of the animal body is due to chemico-physical changes involved in getting the energy from foods, and likewise movements in some lifeless things are due to similar changes. However, such quibbles of advanced physiology have no place in elementary teaching, and for all practical purposes the power of automatic movement is strikingly distinctive of life in general and especially of animal life. Certainly it is the presence or absence of automatic movement that enables us to decide, in most cases, whether an animal is dead or living.

19. *Growth and assimilation.* — The common statement in college text-books, quoting from Huxley, that stones grow by accretion (addition of particles on the outside), while living things grow by intussusception (internal interpolation of particles), is of doubtful use in elementary teaching. It is, also, not strictly true, for there are cases of intussusception in lifeless bodies (Verworn, *General Physiology*, 1897, p. 122). Growth by assimilation deserves more emphasis as a distinguishing characteristic, because this is so closely correlated with the familiar knowledge regarding foods of animals and man. Of course, strictly analyzing, there is no sharp line between growth of inorganic things and that of living things; for, just as the crystal can grow only from the same chemical substance, so the animal grows only from certain chemical compounds — proteins, carbohydrates, fats, and minerals. These four different foods from various sources are possibly identical in the analysis of chemistry, but there is obviously a vast difference in the sum total of their combinations after assimilation. For example, a frog and a bird may eat earthworms, but however similar in chemistry the resulting substances may be, the important fact is that, viewing the result in the living animals, frog substance derived from earthworms is as different from the bird substance derived from earthworms as frogs are different from birds. There is fundamental similarity of chemical composition of frog and bird but, aside from chemical analysis, the two animals are decidedly different. These strikingly obvious differences are what we need for applied biology in elementary courses, leaving the quibbles about the ultimate similarities and differences in chemico-physical structure to the advanced student of pure science. For the purposes of elementary biology, it is sufficient to recognize the general principle illustrated when a frog eats earthworms or plants and assimilates the unlike material to form substance that is obviously frog and not the food that was eaten.

(*D*) If the students have not seen experiments with crystallization, perhaps in the physical nature-study of grammar grades or in introductory physical science, it will be desirable to perform the experiments suggested in the *Applied Biology* and call attention to the main facts that illustrate growth.

20. *Breathing of frog.* — The word respiration is deliberately avoided here. In the present connection it is undesirable to go beyond the most obvious external manifestations of breathing.

Lime-water is made as follows : Slake a lump of freshly burned lime (calcium oxide) with water. Add about a tablespoonful of the freshly slaked lime to a half-pint of water, in a stoppered bottle, and after a day pour off the clear water into another bottle, or filter through coarse filter-paper. Drug stores sell lime-water for use in medicine as a mild alkali, especially used in milk for small children. It can be kept for some time; but before making an experiment, a small quantity should be tested by pouring into a test-tube and breathing into it. A white precipitate should form quickly.

Barium-water, a solution of barium hydrate (costing about 25 cents per pound at chemical supply houses), in place of lime as directed above, is more convenient than lime-water for testing carbon dioxide. A teaspoonful of the barium hydrate will make a quart of the barium-water. Make a solution in cold water in a stoppered bottle, let it stand a day, and then pour the clear water into another bottle. Keep tightly corked, for CO_2 from the air will soon produce a cloudiness due to precipitation of barium carbonate.

Carbon dioxide. — It does not seem desirable at this stage to undertake an extended lesson on carbon dioxide. Simply demonstrate the change caused by breathing, name the substance that is precipitated in the limewater, and leave the discussion of carbon dioxide and its properties until the chemistry of respiration is reached in later lessons.

The explanation that the white precipitate is calcium carbonate in the lime-water, or barium carbonate in the barium-water, formed by a combination of CO_2 and the hydroxide, is interesting to pupils who have studied some chemistry.

The presence of carbon dioxide in the air of the schoolroom may be demonstrated by pouring some barium-water into a shallow glass dish and leaving it exposed to the air for an hour or two. A white film usually forms in a short time.

Control experiments. — The experiment with the frog gives the first good opportunity for a "control," *i.e.*, an exactly parallel experiment without an important factor (in this case a frog), and serving to prove by comparison that the omitted factor is responsible for the result. Such "controls" should be made whenever possible. In § 26 on plant breathing, there is the same need of a "control."

Even such a simple experiment as breathing through a straw into lime-water does not in itself prove that breathed air is changed; we must show that fresh air does not cause a precipitate in lime-water. This may be proved by shaking some lime-water in a jar filled with *fresh* air, or by pumping such air into lime-water with a pair of bellows or an atomizer. The great value in all scientific experiments lies (*a*) in imparting information in the most impressive way known to educators, and (*b*) in leading students to logical thinking. The relative values of these two results are not yet known (see Bigelow, 1904, pp. 244–259); but certainly no opportunity for logical work should be omitted. Hence it is important that the teacher should go over each experiment and explain each step and what it proves. Herein is the great value of "controls," for they alone teach the nature of scientific proof.

In many parts of the *Applied Biology* the authors have been forced by space limitations to omit detailed directions for "control" experiments.

24. *Moving protoplasm.* — Leaflets of Elodea (also known as Anacharis), the staminal hairs of Tradescantia virginiea, the leaflets and internodal regions of Nitella and Chara, and root hairs of rye and other seedlings grown on moist paper may be used for demonstrating movements of protoplasm. In Elodea the large chlorophyll-bodies (chloroplasts) are carried around with the moving protoplasm; but in Nitella and Chara they are stationary in the peripheral layer of protoplasm. Mount leaflets in fresh water, cover, and use about 100 magnification. It may be necessary to examine several leaflets of Elodea before finding very transparent cells, and the Nitella and Chara plants are often incrusted with lime deposits. Also, the movements often cease temporarily when the plant leaflets are first mounted or if subjected to jar or pressure.

Reference to *F*ig. 30 in 1911 edition should read Fig. 28.

25. "P*lant food.*" — The suggestion in this section that soil fertilizers contain some food for plants is open to objection from the viewpoint of strict physiology; but the popular usage of the phrase "plant food" is necessary for this preliminary survey of plant activities. (See § 97 in this *Manual.*)

26. "P*lant breathing.*" — The word "breathe" is popular and not scientific, and is sufficiently accurate for our present lesson. The strict botanist will insist upon the use of "respire"; but this, too, is not without its possibilities of confusion. (See § 51 in this *Manual.*)

(*D*) If the experiment with a potted plant under a bell-jar is performed, a "control" without the plant should be demonstrated. Also, the bottom of the jar should be sealed to a glass plate with vaseline, or set on a dinner-plate containing a half-inch of water. The experiment must be carried on without light (reserve explanation to students of influence of light until §§ 99–102). Jars may be wrapped with black cloth or paper. A wooden box (about 18″ cube), painted inside with lamp-black mixed with turpentine and little ▮no linseed oil, and with large ventilating holes covered ▮ loose curtains of black cloth, is a useful piece of apparatus for experiments requiring absence of light. In above experiment, cover the bell-jar with the box.

Another method is described in § 106 in this *Manual.*

CHAPTER IV

INTRODUCTION TO ANIMAL BIOLOGY

32–64. *Order of study.* — The text of this chapter in the *Applied Biology* has been arranged so that, if the teacher prefers, it may follow the introduction to plant biology (Chapter V). The reasons for such a change in the order of study might be (1) the greater ease of collecting certain plant materials in early autumn, and (2) the claim of relatively few teachers that plants are better for beginning the study of biology.

When functions are to be considered and the course is really more than advanced nature-study, it is probably far better to begin with an animal, because comparison with the human body makes an easier way to the introduction of physiological principles that apply to organisms in general. However, it is desirable that teachers should try each order of beginning the study with classes in different years or school terms.

If the preference of the teacher or availability of materials favor plant study in the autumn, there will be no difficulty in taking up Chapter V after I, II and III, and then later studying IV. See note in § 65 of this *Manual*.

Frog as a type. — In the lessons of Chapter IV the frog should be studied as a type of animal structure and functions, not simply as an amphibian. Hence the teacher should emphasize those points in which the frog illustrates animals in general, or at least the backboned animals. Those who do not appreciate the fact that study of one animal may teach much general zoölogy should read at least the first chapters of Huxley's *The Crayfish as an Introduction to the Study of Zoölogy* — one of the great zoölogical classics.

The teacher who is preparing to present Chapter IV will do well to become familiar with the first part of Parker and Parker's *Practical Zoölogy,* or with Holmes's *Biology of the Frog.*

32. *The value of the frog* for introductory type study is discussed by Bigelow, 1904, pp. 360, 370, and 378; and suggested outlines for such work are given on pp. 386–388 of the same book. The authors of the *Applied Biology* believe that the students ought to know in advance "why the frog is selected for study." (See § 5 in this *Manual.*)

In a short course where there can be dissection of but one animal, it is worth while to make an extra effort to obtain frogs for that work. Even if it is possible to get but one frog for each group of four students, the authors have found that very satisfactory work can be done under these conditions; and rather than omit the frog because not enough are available even for these groups, the authors advise that the pupils use, in connection with the text, specimens formerly dissected by the teacher and preserved. In such an event, it will be well later to let the students dissect a fish, in order to give them some practice in individual laboratory work.

34. *Defense of animal study* has been found useful (1) in order to overcome the natural repugnance of many students to work in dissection, and (2) because it is a splendid opportunity for calling attention to some facts that will tend to make the students have a sane view of the scientific study of animals. In these days of anti-vivisection crusades, led usually by people who forget or have never known the unnecessary killing and cruelty to animals by hunters and others, and who are blissfully ignorant of the results and methods of biological science, it is important that those who study even the elements of biology should become acquainted with facts that will give a sensible attitude towards scientific study and at the same time towards other problems that arise from man's dominion over animals.

35. *Material for study of frog.* — It is best to keep living frogs (see Bigelow, 1904, p. 410), and chloroform them a half-hour before needed for study. The work cannot be completed in one day, and specimens should be preserved in water containing a small amount of carbolic acid, or other antiseptic.

If frogs must be preserved in the autumn, alcohol is better than formalin for females, because the latter causes swelling of oviducts (see note by Bigelow, 1904, p. 408). Denatured or wood alcohol will answer all purposes, if a school does not use enough in its science departments to warrant buying pure alcohol in bonded barrels.

The comparison of frog and human with reference to dorsal, etc. (p. 26), should also state that in modified animals the lower surface is not always ventral, *e.g.*, a flounder lies on its side.

36. *Frog's heart.* — The phrase "the posterior, conical, whitish part of frog's heart is the ventricle" refers to the dead specimen, in which the ventricle is contracted and empty of blood. Of course, it is reddish when full of blood.

Frog's bladder. — A specimen showing the bladder (p. 35 in *Applied Biology*) may be prepared as follows: Select a frog and kill it with chloroform or ether. Carefully cut open the frog's abdominal wall, tie a thread around the large intestine, and then with a medicine-dropper inject water containing some lamp-black, or other pigment, into the cloaca so as to distend the bladder. The specimen may be preserved for years in five per cent formalin solution.

Circulation in capillaries. — For demonstrating this in a tadpole's tail (p. 32 in *Applied Biology*), use a glass plate such as a 3×4 photographic negative that has been cleaned; or, which is much better, use the cover of a Petri dish (commonly used for bacteria, § 245); or a watch-glass with a flat bottom. Lay the tadpole on its side, pour a few drops of water on its tail, cover its head and body with a piece of water-soaked cotton, and adjust the glass on the stage of the microscope. Use low power. As a rule, the tadpole will lie quietly for some time.

If tadpoles are not available, the web of a frog's foot may be used to demonstrate the capillary circulation. Take a thin board about three inches wide and eight inches long, and bore a half-inch hole about two inches from one end. Select a small frog (one with an unpigmented web on a hind foot, if possible); wrap the body with a strip of very wet cloth; fasten the body to the board by means of small rubber bands; catch some loops of threads around the tips of the toes; and draw the threads so as to spread the web over the hole. The threads are most easily fastened by drawing them beneath some slivers previously raised with a knife at the edges of the board; but common pins may be used. Now clamp the board on the stage of the microscope with the hole placed so that light will be reflected through the stretched web. If the frog struggles violently, pour a few drops of sulphuric ether, or chloroform, on a small piece of cotton and place over the nostrils until the animal becomes quiet. The frog should be released and returned to an aquarium as soon as possible, in order not to keep it long in a position that many students will think more uncomfortable to the frog than it probably is.

Teachers of elementary zoölogy, particularly in public schools, should aim to avoid criticism by those who hold extreme views on the question of cruelty to animals.

38. *Study of microscope.* — The author of this *Manual* will gladly loan to any reader who sends a self-addressed envelope nine or ten inches long, a sample set of type-written sheets for a lesson (one hour) on microscopes, which is required of all beginners at Teachers College, Columbia University, before they go on to study tissues. Pictures of microscopes cut from old catalogues and mounted on cardboard, the chief parts being labeled, are used to illustrate the typewritten directions. The leading manufacturers of microscopes will send their customers booklets on how to use the microscope, and some of them will also supply a wall-chart.

38–41. *Preparations* needed for most of this introduction to microscopic study may be made from fresh materials. Some permanent preparations are desirable. Teachers who lack training in microscopic technique will find some general directions in Parker and Parker's *Practical Zoölogy*, pp. 121–125, 135–139. Those wishing to learn the special technique should read Guyer's *Animal Micrology* (University of Chicago Press), and similar books by Gage, Lee, and others. (Listed by Bigelow, 1904, p. 393.)

Preparations may be purchased from dealers named in Appendix III of this *Manual*, and from others listed by Bigelow, 1904, pp. 414–416.

39. *Epidermis of frog.* — Two or three frogs kept for a few hours in a small aquarium or battery-jar will usually rub off large sheets of their epidermis. Collect, wash well, and preserve in strong alcohol. Almost any aniline dye dissolved in alcohol or water will stain the cells. Ordinary red or green ink diluted with water also works well, if special stains are not at hand.

41. Concerning *moving protoplasm* see § 24 in this *Manual.*

42–55. *Beginning physiology.* — These sections in the *Applied Biology* are intended as a general introduction to the elements of animal physiology. The relation of such an approach to human physiology is discussed by Bigelow, 1904, pp. 272–275, 459, 462–464. In such a preliminary survey as that given to the frog in §§ 42–55, it seems best to avoid details; *e.g.,* digestion is described as caused by secretions, which are not named and located in this chapter.

45. *Absorption.* — The experiment with blotting paper is a case of filtration and only suggestive, for osmosis is involved in absorption in the digestive organs of animals. However, the same result might be obtained by osmosis through several layers of parchment or other membrane that allows true osmosis.

51. *Breathing.* — Some authors prefer to limit the word "breathing" to the mechanical or muscular work of pumping air into and out of the lungs, and not as a popular synonym for respiration; but it seems doubtful whether science can compel a strictly scientific usage of such a popular word, and for accurate purposes it is better to use the term "respiration." It is still better, in the opinion of the writer, to recognize that respiration is an old term combining and complicating two processes (viz., oxygen supplying and carbon dioxide excretion); and hence in the *Applied Biology* these are treated separately in §§ 48, 50, 105, 106, 425–431, 433.

Respiration. — In treating oxygen-supply and excretion of carbon dioxide in separate sections, the authors are aware of a rather radical departure from the usual presentation; but many years' trial with both high-school and college classes has led to the conviction that much clearness is gained by avoiding the time-honored confusion involved in considering respiration as one function with two opposed processes. Because the lungs happen to perform two processes at the same time is no logical argument for treating such processes together and thereby confusing them. A human hand may wield a hammer or work a piano, but the doing of two kinds of work by this same organ is no reason why blacksmithing and piano-playing should be studied together.

Of course, the fact that there is a complementary relation between the amount of oxygen absorbed and of carbon dioxide excreted is a reason for considering them together, but this is a problem for advanced study, rather than for the high school.

See note on § 100.

52. *The transporting work of blood* with regard to foods, oxygen (§ 48) and excretions (§ 50), needs frequent emphasis. Of course, lymph in the frog also plays a part in the same functions; and might be described in connection with this section (see § 408 in *Applied Biology*).

56–62. *Frog development.* — As introductory to this lesson, review §§ 21 and 22. It is advisable to have material in formalin for illustration, but living frog's eggs should be followed through their development in early spring. Of course, details of the cleaving egg, formation of germ-layers, etc., do not belong in elementary work.

59. Reference to Figs. A, B, C, D (bottom of p. 59 in *Applied Biology*) should read *F*igs. 22, A, B, C, D.

CHAPTER V

Introduction to Plant Biology

65–110. *Order of study.* — If the course begins in September, some teachers may prefer to make use of plant materials available then. This may be done by assigning this chapter subsequent to Chapters I, II and III, and then either taking Chapters IV, VI, VII and VIII, in this order, or (less desirable) following Chapter V with Chapters VIII and IX. The authors believe that Chapters IV, V, VI and VIII will be found so arranged that any one of these orders of study may be easily followed. See also note in § 32 of this *Manual*.

The terms. "flowering" and "flowerless" are deliberately used in § 65, because at this stage they will convey more meaning to the students than would any other words. This is certainly not the time to introduce beginners to terms such as phanerogams, spermatophytes, and cryptogams, concerning the use of which even the botanists disagree.

65. *Comparative botany for beginners.* — Judging from the majority of text-books of elementary botany and especially from the most successful ones, the approved and highly orthodox way of beginning the scientific study of plants is to make a series of is of seeds, stems, roots, flowers, etc.; and wi of these specimens to plunge a beginner into the man of comparative botany. The student first compares common seeds with seeds of perhaps a half-dozen plants he has never seen; then he is pushed on to make a comparative study (morphological) of roots of unfamiliar plants; and so on through the collections of stems, leaves, flowers and fruits. The final outcome in the mind of the student is supposed to be a sort of composite photograph of the best that it is good to know about plants and their life. This is certainly a result to be desired; but the trouble is that few average students ever put these stray ideas together and thus get a clear conception of the interrelations of the various organs of a plant considered as a living individual. In fact, the students really do not know any one plant, but rather a mass of disjointed pieces of plants. After much critical observation, experimentation and reading of examination papers, the authors of the *Applied Biology* are convinced that few beginners get from such comparative botany as clear an idea of the life of a plant as they often get in the same length of time from the study of animals. The difficulty seems to be connected with the intense comparative study of plant parts before the students have been prepared for such work. Comparing this method of beginning botany with the common order of beginning zoölogy, it is evident that the very nature of animal materials has tended to force well-rounded studies of at least one type of animal, before an attempt at comparing several animals is made. Even a botanist who teaches beginning botany comparatively would consider it absurd for a zoölogist to make collections of heads, legs, hearts, stomachs, etc., and then to proceed to teach beginners by having them spend a week or more in comparing all available legs, then a week on the study of typical heads, hearts, stomachs, and so on in imitation of the seed-root-stem-leaf plan, comparing intensively the various types of each organ before the students know the meaning of that organ in the structure and life of at least one animal. The very suggestion sounds absurd to a zoölogist, and so it is; but it is practically just what-we have long considered proper for beginning botany. Probably the chief reason for the wide adoption of this order in botany is that it is easy to go out and collect a lot of leaves, stems, or flowers; but it is not at all easy to assemble a set of legs, hearts, or stomachs of animals.

Comparative study in beginning botany received a powerful impetus in the earlier teaching from the systematic point of view (*e.g.*, Gray's *Lessons in Botany*), and the convenience of materials has tended to perpetuate the method long after the point of view has changed decidedly.

The arguments (see Bigelow, 1904, pp. 352–355) for general study of an animal type as an introduction to zoölogy are certainly applicable to botany. In fact, the plan has been worked out for a plant in Sedgwick and Wilson's *General Biology*, which in its general outlines must appeal to every one who has the point of view of general biology, even though one may question whether the fern as a type is the best available plant (Lloyd, 1904, p. 114).

The study in Chapter V of the *Applied Biology* is planned to introduce plant biology along lines parallel with those adopted in the previous chapter for animal biology. It will be obvious to the critical reader that in certain cases the detailed treatment has been made from the view point of zoölogy rather than from the viewpoint that prevails in most text-books of clementary botany. As an example may be cited the physiological topics under headings parallel with those applied to the animal side. The writer urges as a justification for this the fact that the greatest interest in the study of biology is, for the average citizen, gained by viewing it from the standpoint of animals (discussed by Bigelow, 1904, p. 252); and therefore in order to develop the greatest value from plant study, the presentation of facts must be harmonized as far as possible with those on the animal side, which in turn have a bearing upon the human aspect of biology.

66–81. *Bean plant for introductory study.* — After much search for a plant adapted to a study parallel with that of the animal in the preceding chapter, the bean plant was selected. It was used by Huxley and Martin in their *Practical Biology*, and is recommended by Ganong, who says (1899, p. 194) that "the advantage of following some one kind of plant through the entire cycle is very great."

66. *Bean plant.* — Materials : young bean plants, some three and some five weeks from beginning of germination, four or five inches high, some growing in pots or boxes. Some may be preserved in fruit-jars with two to five per cent formalin solution. Also have some plants in flower and pod. These can be grown in six or eight weeks by planting some "extra early" variety (see seed catalogues). Also have bean plants growing in pots, or at least museum specimens (in formalin) of such plants, from seeds planted one inch deep and others four inches deep, some three weeks' growth and some six weeks' growth.

Reference : "Beans" (*Farmers' Bulletin 289*).

68. *Roots.* — Flower-pot saucers are excellent for growing roots and root-hairs. See Ganong, 1910, 342, 343.

Strictly speaking, only the cell-walls containing protoplasm and nuclei are complete cells, as defined on pages 40 and 41 ; and the honeycombed appearance in the center of sections of roots and stems is due to empty cell-walls, which formerly contained protoplasm and nuclei.

69. The mucilaginous part of slippery-elm bark belongs to the inner bark outside the cambium (top of p. 71).

70. See note on cells in § 68 above.

73. *Leaf epidermis.* — The purple leaves of the plant known in greenhouses as Tradescantia are excellent and may supplement material from bean leaves.

75. *Bean flowers.* — Eight or ten weeks before this lesson, plant seeds of extra-early beans (*e.g.*, Early Valentine or Early Six-Weeks). In the early autumn, bean plants with belated flowers may usually be found in gardens. When grown in schoolrooms in pots and window-boxes, the plants often flower. Pods ("string beans") may be purchased at markets, or kept preserved in two to five per cent formalin solution.

Ovules and ova. — Avoid confusing plant ovule (p. 78) and animal ovum (egg-cell), for within the ovule is the plant egg-cell.

Similarly, pollen-grains do not correspond to sperm-cells of animals and cryptogamic plants. A correct comparison in popular form is the statement that some protoplasm inside the pollen-grain becomes a fertilizing cell equivalent to a sperm-cell of an animal or lower plant.

Many teachers, some elementary books, and a well-known book on heredity confuse the animal and plant reproductive cells in the way stated above.

77. *Dorsal and ventral.* — The statement that the concave edge of the pod is called ventral because down may prove misleading in that it may suggest that ventral and lower, dorsal and upper, are *always* synonymous. There are some exceptions that should be mentioned. For example, some animals lie constantly on their sides ("flat fishes," or flounders) and hence ventral is not "down," but it is determined by comparing the position of the organs (especially the backbone) with typical fishes. Likewise, the lower surfaces of some parts of plants show by their structure and development that they correspond with the dorsal side of typical cases, and hence lower in such cases is dorsal.

78. *Bean seeds.* — Materials : lima beans, scarlet runner, white Dutch runner, yellow six-weeks, golden-eyed wax, Windsor beans (Vicia). The runner beans named above are not as likely to decay during germination as are limas.

Details of structure. — In the study of the bean seed, such details as chalaza and raphe are best omitted. They mean little except in comparison with embryonic stages that beginning students cannot possibly know or understand.

References for teachers: Lloyd, 1904, p. 146; Ganong, 1899, pp. 161–166. Also see §§ 135–145 in this *Manual.*

Reference for students: Atkinson's *First Studies of Plant Life*, and his *Botany for Schools.* Also the various Bergen botanies.

79. *Bean markings.* — The heart-shaped and translucent marking seen in green beans on the side of the hilum opposite the micropyle is at one end of a ridge (raphe), which is formed from the stalk of the ovule. At the other end of the raphe is the chalaza. The heart-shaped marking has been called strophiole. Of course, these details are not for beginners in biology.

81. *Germination* as applied to awakening of seeds has been in use so long that it seems absurd to insist in elementary courses upon the strict technical meaning as does Coulter in *Plant Structures*, pages 187 and 214. There seems to be no possible confusion of the phrases "germination of spores" and "germination of seeds." At any rate, the long-established popular usage of the word "germination" cannot be changed by any attempt of biologists who prefer to limit it to a technical meaning in connection with spores.

Another method for germinating seeds (p. 85 in *Applied Biology*). — Line a glass tumbler, or a cylindrical lamp-chimney, with blotting-paper (preferably dark colored); fill the center with moist sphagnum, sawdust, soft paper, or cotton; push seeds down between the glass and the blotting-paper; keep moist and warm; shade with a sheet of paper held around the glass by means of a rubber band. A modification of the same method is as follows: Select two sheets of glass of equal size. Lay one on the table; on it place a layer of cotton or several layers of soft papers, then a sheet of dark-colored blotting-paper; on this arrange the seeds about two inches from one edge of the glass; then cover with the other glass; and finally tie the two glasses together or hold with strong rubber bands. Keep moist, and shade with a sheet of paper.

The *growth of molds* in sawdust (p. 85 in *Applied Biology*) often interferes with the germination of seeds; but this may be prevented by steaming the sawdust in a sterilizer or by boiling it in water and then allowing it to drain.

82–86. *Soil.* — References: *Farmers' Bulletin 408;* Goodrich's *First Book of Farming;* Burkett, Stevens and Hill's *Agriculture for Beginners.*

82–122. *Plant physiology.* — Many excellent experiments are described in *Farmers' Bulletin 408.* Osterhout's *Experiments with Plants* is very useful. Atkinson's books, especially *First Studies of Plant Life,* and *Botany for Schools,* Ganong's *Teaching Botanist* and his *Laboratory Course in Plant Physiology* will suggest new ways and additional experiments.

84, 85. *Conserving water in soil.* See *Farmers' Bulletin 266* and *Yearbook Reprint 495,* 1908, on "Soil Mulches for Checking Evaporation."

88. *Osmosis.* — Many teachers demonstrate osmosis by means of hen's egg arranged as follows: Select an egg several weeks old in which the air space at the larger end has been enlarged by evaporation of water (space may be seen when holding egg before a lamp). Carefully chip a small hole so as not to puncture the shell-membrane, or stand the egg on its larger end in a small dish containing an inch of vinegar until mineral matter of shell is dissolved at this end. Now punch a hole in shell and membrane at the small end of the egg, insert a small glass tube 12 to 20 inches long, and pour melted paraffine or sealing wax around the tube so as to seal to the shell. Now stand egg in a tumbler with water about one inch deep and osmosis will begin if the membrane at the larger end has not been punctured.

The experiment suggested in § 88 of the *Applied Biology* may be reversed as follows and will eradicate all notions on the part of students that gravitation is somehow involved in osmosis. If we should fill the diffusion-shell and the glass tube with water and place the shell in molasses at the beginning of the experiment, the water will osmose out of the shell more rapidly than the molasses osmoses into the shell, and the column of water will slowly descend.

Note in the experiment in § 88 that as the molasses slowly osmoses out into the water, the water also becomes a solution of sugar; and as the water becomes heavier with molasses in solution, it osmoses into the molasses in the shell more slowly. This will continue until a given quantity of the mixed liquid outside the diffusion-shell, compared with the same amount of fluid from within the shell, will be found to have equal quantities of molasses and water. In other words, water will osmose into the molasses and molasses into the water until an equilibrium is established.

Preserving membranes. — Adding some formalin, boric acid, menthol, or thymol to the water will prevent fermentation and injury to the membrane by bacteria. As soon as the experiment is completed, wash the membrane, and either insufflate so as to dry quickly, or preserve in a fruit-jar with a weak solution of formalin or other antiseptic.

Popular expressions. — The word "attracts" as a short description of osmotic phenomena (pp. 90, 92 in *Applied*

Biology) is criticized by some physicists. Molecular explanation may be given to students who have had physics, but others must be content with observing the fact that the water osmoses into the molasses faster than the molasses into the water. This statement will avoid the word "attracts," which, like "suction" for air-pressure in pumps, may mislead as an apparent attempt at explaining rather than at describing a phenomenon. However, there appears to be no more reasonable objection to the use of "attracts" and "suction" for the sake of brevity than there is to saying that the "sun rises," the "wind blows," and many similar phrases. Such time-honored popular expressions are exceedingly useful, and they certainly do not mislead any one who studies science, which explains such natural phenomena.

Diffusion-shells, $\frac{1}{2} \times 3$ inches, cost about 15 cents each and $1.50 per dozen. Goldbeater's bags, 10 to 12 cents each, $1 to $1.25 per dozen. Eimer and Amend, New York. In order to help teachers who wish only one or two shells or bags, the author of this "Manual" will mail them at cost to any one who sends a stamped and self-addressed envelope and postage stamps in payment.

Parchment paper tubing may be purchased for about 10 cents per foot. A piece 3 inches long may be soaked in water, folded, and tied at one end so as to form a water-tight cup. See Ganong, 1910, pp. 344–348. See also § 398 in this *Manual*.

Exosmosis. — If students are confused by the fact that sugar osmoses out into the water, which does not happen with living root cells, perform an experiment with a semipermeable membrane, which prevents sugar from osmosing outward. See Ganong, 1910, p. 347.

91. *"Suction."* — The teacher should make sure that the students understand that atmospheric pressure is responsible for the "lifting power of transpiration." Thus understood, there seems to be no reasonable objection to the word "suction" as popularly applied to pumps. See note on "popular expressions" in § 88 in this *Manual*.

92. *Ascending current.* — For demonstrating the rise of colored fluid in stems, catnip twigs are excellent, and may be obtained late in the autumn after many other plants are frosted.

93. *Transpiration* — The word "evaporation" has been used freely because it requires no definition. Transpiration is now known to be evaporation. The word "transpiration," meaning breathing or exhaling, is obviously an heirloom from the old-time botanists, who seem to have considered transpiration of water as part of a breathing process similar to that of animal lungs, because in both watery vapor is given off. However, it is best that beginners should not know the etymological derivation of the word "transpiration," for it suggests connection with the true respiration process. Similarly, it will later be pointed out that in animal physiology there is a decided gain in clearness by considering the oxygen-supply and the excretory phases of respiration as essentially independent and merely associated in the same organs. After the students understand the method by which water is lost from the leaves, it is well that they should learn "transpiration" as a single word meaning "evaporation of water from plant leaves." To reverse this order of really getting the idea before the word is to encourage the all too common practice of using scientific terms without teaching the ideas for which the words should stand.

Reference : Concerning transpiration, see Ganong, 1910, pp. 331–335; and especially in connection with the experiment on page 96 of the *Applied Biology*.

97. *Plant foods.* — In popular usage, "fertilizers" are known as "plant foods"; but some botanists object to this, reserving the word "food" for carbohydrates, fats, etc. "Raw plant food" seems still more misleading as applied to nitrates, carbon dioxide, etc. Lacking an authoritative term, the phrase "plant food-materials" has been used in the *Applied Biology* for the simple compounds required by green plants.

98. *Elements in carbohydrates.* — Strictly speaking, a carbohydrate is composed of C and the hydroxyl OH, which latter is derived from water absorbed by plants (p. 100 in the *Applied Biology*). Students who have studied little chemistry should be instructed that for carbohydrate-making the plant gets three elements, namely, carbon from carbon dioxide and hydrogen and oxygen from water, and that the oxygen of the carbon dioxide is not built into the sugar or the starch that is made.

At the top of page 100 in the *Applied Biology* change "most" to "numerous" so as to read: "Numerous plants without chlorophyll are saprophytes."

99. *Saprophytes.* — Insert the word "non-parasitic" before "plants" in the last clause of the first paragraph, so as to make it mean that non-parasitic plants without chlorophyll absorb food from decaying organisms.

Carbon dioxide taken from the air and used in making starch may be demonstrated as stated by Ganong, 1910, pp. 301–303. A square meter of leaf surface absorbs from air $\frac{3}{4}$ of a liter of pure carbon dioxide per hour in bright daylight, and gives off almost the same amount of oxygen.

100. *Photosynthesis.* — Many teachers speak of "starch-making" as if synonymous with photosynthesis, but that is probably in large part sugar-making. Hence it is "carbohydrate-making." Photosynthesis involves the idea of carbohydrate synthesis by the action of light on chlorophyll-containing cells. There is no good substitute for the word. The process is of such great significance that the term deserves memorizing.

D

Sugar is probably first formed in photosynthesis and then converted by enzyme action into starch; but only experiments dealing with the starch in leaves are practicable in elementary studies.

Starch in leaves. — Some kinds of leaves (*e.g.*, nasturtium) require about forty-eight hours in darkness before the starch entirely disappears.

It has been shown that direct sunlight gives more light than leaves can use in photosynthesis (bottom of p. 103 in the *Applied Biology*).

Starch-test. — While starch is the only *common* substance in plants that gives the deep blue color with iodine, there are other substances that give some lighter shades of blue (*e.g.*, put a drop of iodine solution on filter-paper or cotton and the cellulose is often colored light blue, especially if there is acid present). In all doubtful cases, chemists use special tests in addition to the iodine test.

Starch in leaves. — The directions for covering part of a leaf with cork or tin-foil, *Applied Biology* (p. 103), should have cautioned against fitting either of these substances tightly on the lower surface of a leaf, for the free entrance of carbon dioxide may be prevented thereby. A partial control experiment may be made by pinning corks only on the lower sides of some leaves, and comparing these with leaves covered on both sides. The absence of starch from the leaves with corks on lower sides suggests that carbon dioxide has been kept out of the stomata. See Ganong in *School Science and Mathematics*, Vol. 6, p. 298, April, 1906. Also his *Plant Physiology*, second edition, p. 90; and his *Teaching Botanist*, 1910, pp. 294–297.

101. *Starch disappears* from leaves in light as well as in darkness; but this is not easily proved, because starch is formed more rapidly than it can be removed from leaves. See reference to digestion in § 102 in the *Applied Biology.*

105. *Oxygen 'from plants.* — In demonstrating the liberation of oxygen from Elodea, Cabomba, Spirogyra or other water weed (p. 112), keep the funnel several inches from the bottom of the battery-jar in order to allow the entrance of more carbon dioxide. Select a large battery-jar or aquarium, which exposes a large surface of water to the air. Actively growing plants in warm, sunny weather give best results. Land plants, sometimes suggested for a similar experiment, are useless (Ganong, 1910, pp. 238, 239, 308, *F*ig. 23; also in *School Science and Mathematics*, Vol. 6, 1906, p. 297).

See note in § 100 of this *Manual*.

106. *Method for demonstrating CO_2 excretion.* — *A* represents a pan or dish partly filled with water. *C* is a wooden block or other support for saucer *B*, which holds lime- or barium-water. *D* is the position of a green leaf which may be laid across top of dish *B*. *E* is an inverted bat-

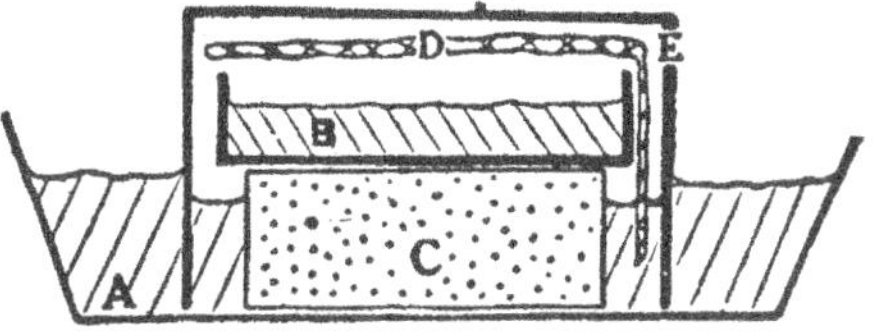

APPARATUS FOR DEMONSTRATING EXCRETION OF CARBON DIOXIDE. (See Text, § 106.)

tery-jar or crystallizing dish. It is well to draw out some water from inside *E* by means of a rubber tube. A film of calcium (or barium) carbonate forms in *B*. A parallel control experiment should omit the leaf. Of course, light should be excluded; but a parallel control kept for the same hours in light is important. (Jean Broadhurst, in *Torreya*, Dec., 1911, p. 261.)

CHAPTERS VI AND VII

COMPARISON AND CLASSIFICATION OF ANIMALS AND PLANTS

114. *Metabolism.* — Substitute the word "nutrition" for "metabolism" in third line from bottom of page 125 in the *Applied Biology*.

At the top of page 126 in the *Applied Biology* is the sentence "This is constructive metabolism." Some botanists hold that synthesis of foods (starch, fats, etc.) in plants is better not included in constructive metabolism, but limit that term to formation of new protoplasm.

122. See note on § 75 in this *Manual*.

127. *Phylum Chordata*, instead of Vertebrata, is now approved by many zoölogists; but see § 343 in the *Applied Biology* and in this *Manual*.

122. *Sperm-cell and pollen-grain.* — On the common confusion of animal sperm-cell with plant pollen-grain, see § 75 in this *Manual*.

123. *Linnæus.* — The statement that he "introduced" binomial nomenclature might be interpreted as "invented"; but the idea of a double name for animals and plants antedates the work of the famous botanist. The fact is that his writings brought it into universal use among biologists.

129. *Homology.* — The word "homologies" for structural resemblances might be introduced in this section of the *Applied Biology*, but it has no particular advantage at this early stage. The abstract idea of "homology" certainly does not belong here.

132. *Plant names.* — In Part II there are few scientific names of plants, but it is suggested that teachers supplement the book by mentioning important generic and sometimes also specific names of familiar plants. This is better than loading the text with such names, for they certainly tend to give some young readers the impression that biology is difficult.

133. *Classification.* — Some authors prefer the word "sub-kingdom" instead of the words "phylum" (for animals) and "division" (for plants).

There are authorities who would add bacteria and slime-molds (Myxomycetes) as sub-divisions 3 and 4 under Thallophyta (p. 144 in *Applied Biology*).

The division of Thallophytes into Algæ and *F*ungi is a physiological rather than a true taxonomic arrangement, but is still the best for elementary study and does not in the least interfere with advanced systematic studies.

It should also be noted that the slime-molds (Myxomycetes or Mycetozoa) are, by some authors, regarded as animals, not plants. (See "Mycetozoa" in new *Encyclopædia Britannica*, 1911.)

Number of known species of animals. — It should interest students to know that there are many recently discovered species of animals. In the following list the figures in parentheses are the estimates of number of species known twenty-five years ago, while the larger figures are 1911 estimates (*Science*, Vol. 35, p. 468, March 22, 1912.)

Insects, 360,000 (200,000). Protozoa, 8000 (4130). "Worms," about 14,000 (6300). Sponges, 2500 (600). Cœlenterates, 4500 (3000). Echinoderms, 4000 (2370). Myriapods, 2000 (800). Arachnids, 16,000 (4000). Crustaceans, 16,000 (5600). Mollusks, 61,000 (21,320). Fishes, 13,000 (9000). Amphibians, 1400 (1000). Reptiles, 3500 (2500). Birds, 13,000 (10,000). Mammals, 3500 (2300). These and smaller groups bring the totals up to 522,400 in 1911; 273,220 in 1886.

CHAPTER VIII

SEED-PLANTS

134. *Order of study.* — In addition to variation in order of study of chapters in the *Applied Biology* as suggested in §§ 32 and 65 in this *Manual*, it would be easy to start with this Chapter VIII and give the usual kind of botany course with comparison of seeds, roots, etc., from the very beginning. If that plan is preferred for plant study as a part of biology, or for a short course in botany, assign the sections of the *Applied Biology* as follows: Begin with bean seed (§§ 78, 79); then study the bean pod (§ 77) enough to show its relation to the seeds; then §§ 80, 81; then other seeds (§§ 135–145). Next study roots (§§ 67, 68, 146–155). Then stems (§§ 69, 70, 156–181). Next leaves (§§ 72, 73, 181–188). Then flowers (§§ 74, 75, 189–211, possibly omitting §§ 209–211). Finally, in the studies of seed-plants take up fruits (§§ 76, 77, 212–218). A series of studies of plant physiology (§§ 82–110) may come at this stage; or those studies may be broken up and functions of roots, stems, leaves, and flowers included in connection with lessons on these topics taken from sections suggested above. For example, §§ 82–90 with roots; 91–98 with stems; 99–106 after leaves.

Studies of spore-plants, selected from Chapter IX, may close such a specially arranged course in botany.

The above suggestion in regard to starting with Chapter VIII is another possible variation in the use of the *Applied Biology*, but the authors of that book would prefer a short course in botany consisting of Chapters I, II, III, V, VII, VIII, and IX, and in this order. This might be shortened by omitting the animal part or even all of Chapter III; by selecting the essentials of plant physiology in Chapter V (this will be done in the briefer book, *Introduction to Biology*) ; and perhaps by omitting Chapter VII.

134–264. *From higher to lower plants* is•the order preferred by the authors of the *Applied Biology.* However, great modification of this order, even complete reversal so that one-celled plants are studied first, might be very successful in the hands of expert teachers of botany Such an order has been used for animals in Part III of the *Applied Biology.* But in all cases a many-celled type should be used in Part I as an introduction.

134. *Emphasis on seed-plants.* — Many botanists whose own researches are chiefly connected with the cryptogamic plants are fond of criticizing those elementary textbooks of botany that give most of their prominent space to studies of seed-plants. There is some justification for this point of view in the pure science of botany; but it cannot be denied that in our practical life the seed-plants are vastly more important than the lower forms. Hence botany taught as a part of general education may properly give very much attention to the seed-plants. Certainly in a course of applied biology the higher plants deserve great prominence. The plain truth is that many of the facts and theories that make study of cryptogamic botany so fascinating to advanced students are far from being illuminating to those who are making a beginning in the study of plant life. (See Lloyd, 1904, p. 119; Ganong, 1899, p. 34.)

In the chapter on seed-plants in the *Applied Biology* the authors have included those facts of structure, function, and development that have been pronounced most interesting and most useful by numerous intelligent citizens who have only general information concerning plants. Such a basis of selection has undoubtedly caused neglect of certain topics that many professional botanists would emphasize, and the insertion of some material that has no direct bearing on the pure science of botany. All this is the result of trying to find that botanical study of the seed-plants which may have practical, intellectual, and æsthetic application in the daily life of average intelligent citizens. (See discussion of "botany for the masses" by Lloyd, 1904, p. 88.)

Since the study of the bean plant (or some other available one) has given general knowledge concerning a seed-plant, it seems best that this chapter should deal with the most interesting and most useful adaptations of common seed-plants. For elementary lessons, and for all practical purposes outside of technical discussions in advanced botany, there appears to be no more convenient arrangement of studies of seed-plants than the familiar one of seed, root, stem, leaf, flower, and fruit. For teaching purposes this presents plants as beginners actually see them and, as a practical matter, the teacher can not do better in a first short course than to treat seed-plants as composed of several definite units, organs, or elements (root, stem, leaf, flower) capable of some adaptive modification within limits. Such an interpretation inevitably arouses the interest of the beginner, whose studies are necessarily comparatively anatomical at basis.

It is true that such a division of seed-plant studies is in danger of drifting far toward an out-of-date, formalistic conception (see Ganong, 1899, p. 144); but there will be a minimum of such danger and a decided gain in efficiency of the teaching if to the old botanical trinity of root, stem, and leaf considered as established organs or units from which all other parts came through modification, there be added buds, flowers, fruits, and seeds, as topics for comparative studies. And within this series of topics it will be helpful to point out similarities, taking care that in doubtful cases similarities are not understood as proving origins; *e.g.*, leaf-like stamens do not prove that stamens are modified leaves (see § 185).

135–145. *Seeds.* — References for students : Atkinson's *First Studies*, and his *Botany* (1910), Chap. I ; Andrew's *Practical Botany;* Bergen's *Essentials* and *Foundations*, Bergen and Caldwell's *Practical Botany* (1911).

For teachers : Ganong (1899 or 1910), Part II ; Lloyd (1904), pp. 146–153. "Testing Seeds" (*Farmers' Bulletin 428*).

135–217. *Seed-plants.* — These sections in the *Applied Biology* direct the comparative study of seed-plants along such familiar lines that few specific references seem needed in this *Manual.* The same topics (root, stem, leaf, etc.) occur in many well-known text-books of botany to which teachers and students are likely to refer.

References : The authors of the *Applied Biology* would commend the books of botany for high schools by Atkinson, Bergen, Bergen and Caldwell, Bailey, Stevens, Andrews, Coulter, Leavitt (revised Gray), Osterhout, Clute, Bessey, Barnes for reference or supplementary work on most points mentioned in Chapter VIII of the *Applied Biology.* (See selected list in Appendix III of this *Manual.*)

An up-to-date list of advanced books of botany for teachers is given by Ganong, 1910, pp. 176–226 — a chapter that every teacher of biology should read.

137. *Squash seeds.* — Most varieties of squashes and pumpkins have seeds too small to be convenient for study, and seedsmen should be asked to send seeds of the Hubbard varieties or of others equally large.

There is no accurate botanical distinction between the words "squash" and "pumpkin"; *e.g.*, the Connecticut field or "Yankee" pumpkin, the yellow crookneck squash, the scallop squash, and many gourds are varieties of Curcubita Pepo, while the Hubbard and the mammoth squashes (also commonly called pumpkins) belong to C. maxima.

138. *Castor-oil seed.* — It does not seem desirable to confuse elementary study with the details of such markings as raphe, chalaza, etc., because they are of no particular significance to a beginner. In order to answer possible questions, the teacher may need to know some of the details mentioned by Lloyd, 1904, p. 151. The essential points are as follows: The micropyle is surrounded by a spongy mass (caruncle) of unknown function, but possibly for absorbing water when germination begins. This spongy mass is somewhat nearer the side of the seed on which is a longitudinal ridge (raphe). At one end of this ridge (toward the micropyle) is the hilum (best seen in seeds in a mature fruit of the castor-oil plant), and at the other end of the ridge is the chalaza. After removing the seed-coat, a reddish-colored spot will be seen beneath the chalaza, and from it delicate veins radiate. These veins carried food during the development of the seed.

Supplementary facts. — The authors of the *Applied Biology* have long believed that it is well worth while to make occasional digression and give students supplementary information. Here is such an opportunity: The name "castor-oil plant" refers to the oil obtained by pressing the seeds and used in medicine and for lubricating machinery. The word "castor" probably refers to odor of the plants, which resembles that of beavers that belong to the genus Castor. The botanical name for the plant is Ricinus, from Latin meaning a tick; and was probably given because the seeds resemble wood-ticks or sheep-ticks. The seeds contain a powerful poison known as ricin.

It is true that such facts are not necessary for technical botany; but teachers of science in high schools should not be overanxious concerning the aims and needs of technical science in colleges. The fact is that the questions of students indicate that they are interested in points that are of human interest apart from their scientific connections. Also, occasional reference to the language derivation will add interest to the study both of science and of language. Teachers of science should occasionally take such opportunities for giving a glimpse of the origin of some scientific names, for the sake of broadening interest in language study and its applications.

139. *Corn grain.* — While, strictly speaking, this is a fruit corresponding to a bean pod with one seed, beginners will use the word "seed" in the popular sense; and "seed corn" will continue to be sold and planted in spite of criticism by botanists. The exact meaning of a seed cannot be clear until after the study of fruits (§§ 212–218). The term "testa" is not properly applicable to the corn grain, because its "hull" is a testa consolidated with the wall of the ovary. In fact, the hull is like an imaginary bean pod with one seed whose coat is consolidated with the pod. The sunflower fruit (commonly called "seed") has its testa and ovary wall separate.

The shield-like body (scutellum) embedded in the corn grain is *probably* (not certainly) a cotyledon.

References for teachers: Lloyd, 1904, pp. 149–150; Ganong, 1899, p. 169, or 1910.

References for students: "School Lessons on Corn" (*Farmers' Bulletin 409*); "Germination of Seed Corn" (*Farmers' Bulletin 253*); "Seed Corn" (*Farmers' Bulletin 415*). Also text-books mentioned in § 135 in this *Manual.*

143, 144. *Epicotyl and hypocotyl* are just as easily learned as caulicle or radicle of the older botanists. The word plumule for the bud on the epicotyl is still useful at times. Some authors use hypocotyl, but retain plumule instead of epicotyl.

145. *Physiology of germinating seeds.* — References: Osterhout's *Experiments with Plants*, Chapters I and II. See also Atkinson's books, and those by Bergen.

156, 157. *Twigs and buds.* — Ganong, 1910, pp. 297–300. *Cornell Nature-Study Leaflets.* Comstock's *Handbook of Nature-Study*.

161. *Growth of monocotyledonous stems.* — Bamboo stems grow in height from one to three feet a day and a stem twenty feet high may grow in fifty days, but will be soft and brittle like asparagus shoots, and requires three to four years to harden. The largest Japanese species grows fifty feet high, and five or six inches in diameter.

166. *Grafting* should be demonstrated on seedling apple trees, which may be taken to the schoolroom, grafted, and then planted; or specimens showing various stages (including growth after grafting) should be preserved dry or in formalin.

Reference· "Top-Working Orchard Trees," *Yearbook Reprint 266*, 1902, good account of grafting.

167. *Pruning.* — Small specimens of wild seedling apple trees or other useless trees may be taken to the schoolroom for demonstrating principles of pruning. Specimens taken a year after pruning and showing healing are useful. Reference:. "Pruning," (*Farmers' Bulletin 181*).

169. *Paragraph 3.* — Correct spelling is *lenticels*.

170–178. *Stems.* — Examples of special stem adaptations should be collected and preserved dry or in formalin solution, for use in illustrating the text-book.

180. *Forestry.* — References: "Forestry in Nature-Study" (*Farmers' Bulletin 468*); Roth's *First Book of Forestry* (Ginn and Co.); Pinchot's *Primer of Forestry* (*Farmers'*

Bulletins 173 and *358*, also for sale cloth bound at 35 cents each). Forest Service Circulars: 69 ("Planting"), 130 ("*Forestry* in Public Schools"), 157 ("Conservation"). See list of pamphlets in Circular 19 on *Publications for Use of Teachers* (U.S. Dept. Agriculture).

185. *Flowers and leaves.* — Note that the text of the *Applied Biology* reads that certain parts of the flower are "leaf-like." Strictly speaking, petals and sepals may be regarded as modified leaves, but stamens and pistils are probably not homologous with or derived from green leaves. See Ganong, 1899, pp. 147, 236; or 1910, pp. 230, 352.

188. *Clover leaves* are, from the beginner's standpoint, palmately compound; but in some species the central leaflet is stalked so as to make a pinnate trifoliate leaf. Some authors describe them as "digitate."

189. *Fertilization and embryology.* — This topic has been presented in the *Applied Biology* stripped of all the unessential technicalities which appeal to a plant embryologist or to a cytologist. Elementary botany for high-school education is a questionable place for introduction of nucellus, embryo-sac, chalaza, micropyle, synergidæ, polar nuclei, antipodal cells, secondary nucleus of the embryo sac, pro-embryo, suspensor, hypophysis, and other embryological details; not to mention chromosomes, reducing divisions and similar cytological points which may interest the very advanced student.

With regard to the theory of alternation of generations of angiosperms, it is all very interesting to advanced students in college classes in botany to trace this splendid series of homologies; but it is nonsense to attempt, with high-school classes, to homologize pollen-grain and microspore, embryo-sac and megaspore, ovule and macrosporangium, male gametophyte with contents of pollen-tube, female gametophyte with contents of embryo-sac, endosperm and prothallium.

Double fertilization. — It is certainly best in elementary work to omit reference to endospermic fertilization. As is well known, the antheridial cell in a ripe pollen-grain forms two sperm-cells, which move down the pollen-tube. The nucleus of one of these unites with the egg-nucleus in fertilization, while the nucleus of the other sperm-cell has in *some* plants been seen to unite with the two nuclei in the embryo-sac, thus producing endospermic fertilization. To make this clear to students means unwarranted detail, which will certainly tend to confuse beginners. The text of the *Applied Biology* has been deliberately worded to avoid mentioning that there are two sperm-cells or fertilizing cells in the pollen tube; and it is simply stated that a cell fertilizes the egg-cell. Thus elementary study may escape the complicated problems of endospermic fertilization, which may safely be left for college botany. Moreover, this double fertilization is not known to occur in numerous species, and elementary biology should center attention on the typical aspects of fertilization.

189. *Stamens and pistils as "male" and "female organs."* — The older botanical books refer to stamens as the "male" reproductive organs and the pistils as the "female." This usage is now criticized because comparative botany has made clear that the stamen is a microsporophyll and the pistil is composed of one or more carpels or megasporophylls. Likewise, each anther is composed of microsporangia (pollen-sacs); and each carpel bears megasporangia (ovules). Within each pollen-grain (microspore) is the antheridium-cell that produces the generative or sperm-cell. Within each ovule there is the embryo-sac or megaspore. This forms the archegonium of the gymnosperms and the corresponding egg-apparatus of the angiosperms, and these are the organs for producing egg-cells. Obviously, in terms of comparative botany, the antheridium-cell in the pollen-grain and the archegonium or egg-apparatus in the ovary, represent the real sex-organs; and the male and female reproduc tive organs are hidden within the stamens and pistils.

Such are the detailed facts from the point of view of comparative botany; and beyond question they should be so taught to *advanced* students of college botany. But some recent authors have attempted to transplant this modern plant morphology to the high school, and we read in text-books that "stamens are not male organs and pistils are not female organs," and "pollen-grains are asexual spores," and "the carpels with ovules do not represent female sex-organs." All this is true, but absurd from the standpoint of 99 per cent of high-school students. It is plant morphology run amuck in general education. What does the average intelligent citizen care concerning the homologues of the antheridium or the archegonium? Certainly those interested only in the general results of science will continue to regard stamens and pistils as sex-organs because they are the obvious organs that produce the male and female generative cells. Plant morphologists will not soon convert the general student to the strict botanical terminology, for only those who have the advantage of an extensive comparative course in botany can possibly understand why the ovary and anther of spermaphytes are not in the strict botanical sense sex-organs, but that these are developed somewhere inside of what is called the "ovary." Such quibbles about intricate phylogenetic homologies are quite beyond the understanding and interests of average students. This kind of scientific hair-splitting is well adapted to making science study mysterious, useless in daily life, and consequently unpopular. What the average citizen of general culture will remember after the most strict insistence upon terminology is that ovaries in both plants and animals are organs for producing egg-cells or female reproductive cells, and that the organs variously termed spermaries, anthers, etc., are organs for producing male reproductive cells (sperm-cells). These are the *essential* facts, and the intricate details of histological structure and its phylogenetic interpretation in both plants and animals may well be left for the advanced student, who may reasonably be expected to comprehend the homologies.

However, it is possible in an elementary course to make intelligible the essential facts of reproduction in seed-plants and yet partly to avoid the scientific quibble concerning sex-organs. Simply call the stamens and pistils reproductive organs, which is *practically* but not phylogenetically true, and use the terms referring to sex only when describing the essential nature of the fertilization process — a union of two sex-cells, one male and one female in origin. This is the one big idea in the whole matter; and reasonable high-school science does not need a discussion of why stamens and pistils are not the real sex-organs in modern botany. Too many teachers fresh from university laboratories have busied themselves emphasizing this criticism of the older botany, instead of teaching the essential facts that the average 'cultured citizen needs. The rare high-school student who decides to specialize in biology will have abundant time for learning in advanced college courses the intricacies of sex-organ homologies.

196. *Tubular flowers.* — The modern view (Ganong, 1899, pp. 148, 237; or 1910, p. 235) that the tube of a tubular calyx or corolla is not a union of sepals or petals, but a ring of new tissue surmounted by the free sepals or petals, means absolutely nothing to young students and to those unfamiliar with morphological development of plant structure. It is not advisable to confuse elementary work by attempting such explanations. The observable fact is that the sepals and petals *do appear* united, and it is morphological quibbling to try to inform beginners that the parts are not united but that a tubular ring grew beneath the true petals or sepals.

202. *Ovary inferior.*—See note on § 185. Also Ganong, 1899, p. 148, or 1910, p. 235, on the "calyx-adnate" theory.

207. In legend of *F*ig. 67 transpose words to read "*F*lowers of the composite Arnica."

208. References: "Annual *F*lowering Plants" (*Farmers' Bulletin 195*). "Beautifying Home Grounds" (*Farmers' Bulletin 185*).

212–216. *Fruits.* — The authors of the *Applied Biology* have departed somewhat from the usual classification of fruits, because they have found the subject most interesting under the headings: simple dry fruits, simple fleshy fruits, stone-fruits, complex fleshy fruits; applying the adjective simple to those derived from ovary, and complex to combinations of ovary and surrounding parts.

Educational experiments are needed to determine methods of making study of fruits interesting. At present, it is certainly a topic in botany that competes with bones in anatomy. Probably the greatest interest of students will be found if fruits are studied in relation to flowers and along economic lines. *F*ortunately, all materials needed to illustrate types of fruits may be obtained from markets, or preserved dry and in formalin solution.

Preserve in 2 to 5 per cent solution of formalin various stages from flower to perfectly-shaped fruits of apple, quince, strawberry, buttercup, tomato, gooseberry, cherry or peach, walnut or hickory-nut, raspberry, blackberry, cucumber or squash, sunflower, buckwheat, oak, chestnut burr, horse-chestnut, or buckeye. A series of stages mounted as suggested by Lloyd, 1904, p. 226, are desirable.

See Ganong, 1910, pp. 377–380.

E

216. *Inferior ovary.* — For the interpretation of surrounding receptacle and not adnate calyx, see Ganong,
1899, bottom of p. 147, or 1910, p. 235.

217. *Seed-dispersal.* — Reference for teachers : Ganong,
1910, 272–276.

218. Reference : "Propagation of Plants" (*Farmers'
Bulletin 157*).

CHAPTER IX

Spore-Plants

223. *Ferns.* — In the opinion of many biologists, ferns do not deserve more time in a year's course than that required for §§ 223–229. Teachers who want to provide for more extensive study are certainly familiar with the many excellent text-books of botany such as Atkinson's *Botany for Schools*, Bergen's *Foundations*, Bergen and Caldwell's *Practical Botany*, Clute's *Laboratory Botany*. Sedgwick and Wilson's *General Biology* has an extensive account of Pteris in all its biological aspects.

224. *Fern prothallia* may be found at most commercial greenhouses, but may be grown in the schoolroom. A shallow flower-pot with a mixture of garden soil, leaf-mold and sand should be heated in a sterilizer (p. 256 in *Applied Biology*) for a half-hour to destroy organisms that might interfere with the young ferns. Level and pack the surface of the soil, wet with boiled water, and scatter but do not cover spores. (Spores may be obtained from seed-dealers or collected from leaves placed in paper bags when sporangia begin to be light brown and dry. Spores will keep for years in air-tight jars.) Cover the pot so as to protect against evaporation. Keep the soil moist without sprinkling its surface for several weeks, but set the pot in a saucer with water. Complete dircetions are in Bailey's *Cyclopedia of Horticulture*, which many libraries and most up-to-date gardeners own.

228. *Allies of ferns.* - For supplementary work see the text-books by Atkinson, Bergen and Caldwell, and Coulter. A year's course will not allow time for more than brief demonstrations of specimens selected to illustrate the chief types.

229. *Ferns in coal.* — Most fern-like fossils of the coal are "seed-ferns" (Cycadofilicales) which combine characteristics of the ferns and gymnosperms. The equisetales, the lycopods, and other fern allies are very abundant in coal.

230–234. *Moss.* — References to same authors as for ferns (§§ 223–228).

235. *Algæ and Fungi.* — The simplest possible classification of lower cryptogamic plants has been used in the *Applied Biology* because the division of Thallophyta into the classes Chlorophyceæ, Charophyceæ, etc., of the Algæ; and Phycomycetes, etc., of the true *F*ungi is of no value to one who has not time for thorough study of types of the ten or eleven classes of Thallophytes. Such a sub-division is as useless to the average intelligent citizen as is the detailed classification of the "worms" and crustacea in zoölogy.

237–239. *Unicellular plants.* — Laboratory work on these sections should be illustrative of the text in the *Applied Biology.*

238. *Volvox, F*ig. 95, § 279, in the *Applied Biology,* should be mentioned here as composed of numerous individuals, each similar to Sphærella, set in the surface of a hollow, transparent sphere. It is a colony of individuals, probably plants.

239. *Necessary elements from compounds.* — Page 250, ninth line, in the *Applied Biology,* should read "absorb *compounds* containing the necessary elements (N, S, P, K, etc.)." Of course, not the pure elements, as misleadingly suggested by the omission of the word "compounds."

244. *Molds.* — For the laboratory work in this section, moist slices of bread should be placed in covered dishes (*e.g.*, in glass tumblers covered with 3 × 4 plates of glass), some one and some three weeks in advance, and spores from a piece of moldy bread scattered over the surface. Protect against bright light. Larger pieces may be laid on moist blotting-paper on a dinner-plate, and covered with another plate inverted, or with a battery- or bell-jar. Cultures may easily be made almost pure by following the directions on p. 256 of the *Applied Biology*.

When tumblers are arranged as suggested above, the 3×4 glasses used as covers will often have numerous spores attached to the lower surface and in various stages of germination. Examine such a glass with low power of microscope.

Spores may be germinated in drops of diluted sugar sirup on glass object-slides. Keep the slides in a covered dish along with wet cotton or paper to guard against drying. Do not put on cover-glass until ready to study with high-power lenses.

246. *Ergot.* — On third line of p. 262 in the *Applied Biology* change to read "A fungus of rye." Ergot is often called "smut"; but it is better to limit that word to corn smut and other species of Ustilago.

Economic fungi. — References: "Fungicides" (*Farmers' Bulletin 243*), "Rusts of Grains" (*Bulletin 216, Bureau of Plant Industry*), Weed's *Farm Friends and Foes*, Part III (Heath and Co.).

249. *Mushrooms.* — Reference: "Cultivation of Mushrooms" (*Farmers' Bulletin 204*). Other references are given on page 267 of the *Applied Biology*.

251. *Pasteur's solution* contains the following ingredients:

	PER CENT
Water, H_2O	83.76
Cane sugar, $C_{12}H_{22}O_{11}$	15.00
Ammonium tartrate $(NH_4)_2C_4H_4O_6$	1.00
Potassium phosphate, K_3PO_4	0.20
Calcium phosphate, $Ca_3(PO_4)_2$	0.02
Magnesium sulphate, $MgSO_4$	0.02
	100.00

The author has often found it inconvenient to make up the Pasteur's solution when needed and he now keeps in stock some small packages containing all the constituents (except sugar) mixed ready for adding water. A package to make one quart of the solution will be mailed by the author, only to teachers who use the *Applied Biology*, for twelve cents, actual cost of chemicals, preparation, and postage.

Vacuoles. — In legend of Fig. 83, *Applied Biology*, 1911, add " clear centers contain fluid."

Yeast. — The experiments outlined in this section of the *Applied Biology* are usually impracticable for individual work in high schools with large classes, and the possibilities of error are so great that the demonstration method is preferable. The teacher should have the pupils take notes as the demonstrations proceed. After the facts are recorded, go back to compare the tubes and to draw conclusions. The facts are for the seven tubes as follows: No. 1 (pure water), no growth, no fermentation; No. 2 (pure sugar solution), no growth, some fermentation; No. 3 (sugar with meat extract), much growth, much fermentation; Nos. 4 (crude molasses), 5 (flour paste), 7 (Pasteur's solution), almost the same as No. 3 in growth and fermentation; No. 6 (Pasteur's solution without sugar), growth of yeast, but no fermentation. Conclusions are: No. 1, no food for growth and no sugar for fermentation; No. 2, sugar for fermentation, but not sufficient food for growth; No. 3, the beef-extract evidently adds to No. 2 material for growth, making more yeast-plants and consequently more fermentation of the sugar; No. 4, dark-colored molasses evidently has something for growth in addition to sugar used in No. 2. The sugar used for No. 2 is purified or refined, various substances in the crude sugar being removed. A solution of very dark-brown sugar will give results similar to the molasses. Some substances in the molasses evidently serve the same as the meat-extract in No. 3. Tube 5 shows that yeast can grow in flour paste and ferment it. The explanation is that an enzyme changes some of the flour to sugar and then sugar fermentation by another enzyme occurs. Help students to apply this experiment to bread-making by the usual process. Tube 6 has a solution of known chemical composition (see § 251 in this *Manual*). Pasteur found that yeast must have for growth carbon, hydrogen, oxygen, nitrogen, phosphorus, calcium, magnesium, potassium, and sulphur; and the Pasteur's solution without sugar furnishes these, but there is no sugar to ferment. Tube 7 furnishes the elements needed for growth and also the sugar to ferment.

General conclusions: Yeast may grow in solutions without sugar, but sugar (or starch convertible into sugar) is necessary for fermentation. Yeast cannot grow in pure sugar solution, because sugar supplies only the elements carbon, hydrogen, and oxygen, and six others are necessary.

References for students: Conn, *Bacteria, Yeasts, and Molds*, 1903, chapters on yeast.

References for teachers: Bigelow, 1904, pp. 303–305; Parker's *Biology*, chapter on yeast; Sedgwick and Wilson's *Biology*, lesson on yeast.

253. Reference: "Bread Making" (*Farmers' Bulletin 389*).

255. *Gelatin media* (p. 280 in the *Applied Biology*). Take any commercial meat bouillon or a "clear soup" sold in ten-cent cans, add a solution of sodium hydrate or baking soda gradually to make the bouillon slightly alkaline as shown by bluing of red litmus. To one fourth pint of the bouillon add an ounce of gelatin; and heat and stir in a water-bath until gelatin is dissolved. (An extemporized water-bath may be made on the principle of a double cereal-cooker, by placing a small tin can holding the gelatin inside a larger one containing water.) If the mixture is acid, add more baking soda or sodium hydrate until slightly alkaline. It is now called nutrient gelatin. It may be stored in a cotton-stoppered bottle, sterilized 30 minutes; or it may be distributed at once into test-tubes, flat bottles, or Petri dishes, and then sterilized. Resterilize after a day.

A clearer gelatin may be obtained by clarifying with egg-albumen and then filtering through a hot-water funnel. (See manuals of bacteriology such as those by Moore, Gorham, Heinemann, Stitt; and also some larger text-books, such as Jordan's.)

Instead of sterilizing media in Petri dishes, it is a very common practice to sterilize the dishes thoroughly and later to raise the covers and pour in sterile melted gelatin from test-tube or a flask which has been kept stoppered with cotton.

Some commercial laboratories (*e.g.*, Parke, Davis & Co. of Detroit) sell gelatin media in cotton-stoppered test-tubes and in sterilized bottles; but the cost is too great for class work, being about $2 for a 500 cc. flask.

For very accurate study of bacterial growth, gelatin medium should be made according to directions in bacteriological books, such as those named above.

256. *Disinfectants.*—Reference: "Some Common Disinfectants" (*Farmers' Bulletin 345*). Chapter 17, Conn's *Bacteria, Yeasts, and Molds in the Home*, 1903.

Antiseptics.—The epoch-making work of Lister, who applied antiseptic methods to surgery, should be noted in this lesson.

258. *Soil bacteria.*—References: "Leguminous Crops" (*Farmers' Bulletin 278*). "Alfalfa" (*Farmers' Bulletin 339*). "Cowpeas" (*Farmers' Bulletin 318*). "Soy Beans" (*Farmers' Bulletin 372*). "Red Clover" (*Farmers' Bulle-*

tin 455). "Canadian Peas" (*Farmers' Bulletin 224*).
"Soil Bacteria" (*Yearbook Separate 507*, 1909). "Ni-
trogen-gathering Plants" (*Yearbook Separate 530*, 1910).

258. *Milk and bacteria.* — References: "Bacteria in
Milk" (*Farmers' Bulletin 348*); "Care of Milk" (*Farmers'
Bulletin 413*). Also see Conn, 1903.

259. *Germ diseases.* — References: "Tuberculosis"
(*Farmers' Bulletin 473*). "Typhoid" (*Farmers' Bulletin
478*). Conn, 1903. Sternberg's *Infection and Immunity*
(Putnams). See also § 265 below.

263. *Antitoxins.* — Emphasize the fact that no diphtheretic bacteria, but only their toxins, are injected into
the horse selected for producing antitoxins. The total
amount of diphtheretic toxins injected into the horse,
at intervals of three to seven days for several months,
is enough to kill five hundred thousand guinea pigs,
weighing about a half-pound each, and one cc. of the
immunized horse's blood will contain enough antitoxin
to neutralize the amount of toxin that would kill fifty
thousand guinea pigs.

265. *Books on bacteria.* — In addition to the books
suggested at the bottom of p. 297, in the *Applied Biology,*
Marshall's *Microbiology* has a good general account of
bacteria. Newman's *Bacteria* is good, but needs revision.
See list cited by Bigelow, 1904, p. 423. *Farmers' Bulletin
449* is on hydrophobia.

CHAPTER X

266–367. *Order of study.* — The authors of the *Applied Biology* recognize that many teachers might have more success if the vertebrates followed the earlier frog study directly, instead of this chapter on one-celled animals. The arguments by Bigelow, 1904, pp. 345–352, in favor of beginning zoölogy by studying complex before simple animal types do not, of course, apply to Chapter X of the *Applied Biology*, because the frog study of Chapter IV has preceded as an introductory study.

266–279. Pr*otozoa.* — Concerning the difficulties of beginning work with these animals, see Bigelow, 1904, pp. 345–352. Comparison with complex animals, pp. 379, 380. Materials and methods, pp. 363, 394–397.

References on Protozoa will not be so useful as those on other groups, for the *Applied Biology* probably will require as much time on those animals as can be assigned in most high schools. However, Linville and Kelly, 1906, pp. 280–291; Jordan and Heath, 1902, 11–18; Davenport, 1911, 280–288; and Kellogg, 1901, 81–89; will be satisfactory for students. Teachers will find chapters on Protozoa in Parker's *Elementary Biology*, Parker and Parker's *Practical Zoölogy*, Sedgwick and Wilson's *General Biology*, as well as in all zoölogies. Calkin's *Protozoölogy* is more useful than his earlier *The* Pr*otozoa.*

266. *Frog and protozoans.* — It may be said that after study of plant life in Chapters V–VIII the students have forgotten much of their study of the frog in earlier chapters.

This is probably true, but it is of little significance. Students in the second half-year of any course forget much they learned in the first half; but the efficient teacher has little difficulty in reviewing and recalling the essential facts which may have become more or less dormant in memory.

One strong argument for not going directly from the frog to the other types of animals is that the interval allows the students time to assimilate many essential ideas and to forget more or less trivial facts; and when the great' principles of zoölogy that the frog illustrates are brought up afresh in connection with other animal types, a fixed and definite impression will probably be made.

267. *Paramecium.* — In order to use the high power of the microscope for the study of details of structure, it is necessary to impede locomotion of the paramecia by shreds of cotton, or by adding a mucilage made of gelatin or gum arabic to the water (see directions given by Bigelow, 1904, p. 396).

For other practical points concerning study of Paramecium, see Bigelow, 1904, pp. 363, 379, 395–397.

Conjugation. — In elementary study, it is certainly best not to mention the details of nuclear activity in this process. Dr. Woodruff, of Yale University, has obtained more than 2900 generations of Paramecium without conjugation occurring. Hay-infusion cultures, kept in fruit-jars and examined every few days, will probably develop an "epidemic of conjugation" within a month or two.

270. *Amœba.* — According to Professor A. H. Cole, of the Chicago Normal School, amoebas may be reared in a culture fluid such as Sach's solution. Tablets for making this may be obtained for ten cents from the Agassiz Association, Sound Beach, Conn. The tablets should be dissolved in pond water and various water plants added. See details in Cole's *Manual of Biological Projection* (Neeves Co., Sixty-third St., Chicago).

When referring to the genus Amœba, a capital is proper, but "an amœba" or "amœbas" are used by good authorities when referring to individuals. Likewise, we write Hydra (§ 282) for the genus, but hydras for individuals; and, among plants, Viola the genus and violet the species, Chrysanthemum the genus, and chrysanthemums the varieties or individuals. For convenience, the same anglicizing of numerous botanical and zoölogi cal taxonomic names is needed.

On materials and methods see also Bigelow, 1904, pp. 379, 380, 394. Good permanent preparation of amœbas may be purchased at 50 cents a slide from *F*. D. Barker, Station A, Lincoln, Neb. The chief value of such slides is in showing nucleus. Amœbas are most interesting because of protoplasmic movements. Hence living specimens are important. The same dealer has sent them alive by mail to New York City.

278. *Vorticella allies.* — Examples of colonial Vorticella-like protozoans are described in Parker's *Elementary Biology* and in Parker and Haswell's *Zoölogy*.

CHAPTER XI

Porifera and Cœlenterata

281. *F*resh-water sponges (Spongilla) should be exhibited, if possible, but are too complicated for elementary study. Also, specimens of glass sponges will add interest.

References: Linville and Kelly, 1906, pp. 273–279; Kellogg, 1901, pp. 90–97.

282. *Hydra.* — References for teacher: Bigelow, 1904, pp. 363, 381, 397–400; Parker's *Elementary Biology.*

282–292. *Cœlenterates.* — Concerning selection of types, see Bigelow, 1904, pp. 363, 364. Outline of study, pp. 381, 382. Materials, pp. 397–400.

References for students: Linville and Kelly, 1906, pp. 252–272; Jordan and Heath, 1902, pp. 29–43; Kellogg, 1901, pp. 98–113; Davenport, 1911, pp. 260–279.

For full bibliography of zoölogical books, cited by author and date, see pp. 90–101 in this *Manual.*

CHAPTERS XII AND XIII

"Worms" and Echinoderms

293. *"Worms."* — Certain representatives of that vast and heterogeneous assemblage of animals once involved in the now obsolete phylum Vermes are of such great importance that biology for general education must include them. And yet they are most puzzling in elementary zoölogy, for the fact that, in popular language, they are "wormlike" is the only thing that groups together for beginners the utterly diverse types platyhelminthes, nemathelminths, and annelids. Hence, examples of "worms" should be studied by beginners simply from the point of view of applied biology; and we must frankly admit that their comparative morphology and taxonomy are quite beyond any but advanced students.

The author believes that the popular names of worms (flat worms, round worms, and segmented worms) are sufficient for elementary applied biology; but students of classical languages may be interested in the three technical names that are literally translated flat worms, thread worms, and ringed worms, also in the term helminthology from helminth, meaning a worm. The references given below are general accounts most useful at the close of the study of "worms."

References for students: Davenport, 1900, pp. 151–158; Davenport, 1911, pp. 191–197; Linville and Kelly, 1906, pp. 222–235; Jordan and Heath, 1902, pp. 44–71; Kellogg, 1901, pp. 133–150.

297. *Trichina.* — Reference · "Trichinosis" (*Circular 108, Bureau of Animal Industry*).

300. *Annelids.* — While metameres are mentioned in the *Applied Biology*, the author questions whether a discussion of the principle of metamerism should be introduced to high-school students, even after comparative study of various animals. Merely the fact that earthworms, crustacea, insects, etc., are segmented should be emphasized. The synonym somite for metamere is preferred by some zoölogists.

301. *Nereis* is interesting for comparison with the earthworm, which in adaptation to burrowing in soil has not prominent appendages like those of the sandworm.

References for students : Davenport, 1900, pp. 145–151, or 1911, pp. 182–189; Linville and Kelly, 1906, pp. 222 223.

302. *Earthworm.* — Materials : Bigelow, 1904, pp. 400–404. Its value as a type, same author, p. 360.

There should be only as much study of the internal organs of the earthworm as is needed for appreciation of the general physiological processes. All details, such as locating organs in numbered segments, may well be omitted from elementary zoölogy

A lesson on the physiology of the earthworm may be added if time allows. See Linville and Kelly, 1906, pp. 196–208; or (for teachers only) Sedgwick and Wilson. Ecology of earthworm : Linville and Kelly, 1906, pp. 218–221.

References for students : Davenport, 1900, pp. 133–136; Davenport, 1911, pp. 168–172; Kellogg, 1901, pp. 133–144; Linville and Kelly, 1906, pp. 195–221. At close of lesson on earthworms, Chapter XI (pp. 161–167) in Davenport, 1911, may be assigned for reading, or outlined by the teacher.

Reference for teachers : Sedgwick and Wilson's *General Biology.*

303. *Development of earthworm.* — For teacher, Sedgwick and Wilson, or Linville and Kelly (pp. 209–217) give the essential facts and diagrams.

305. *Echinoderms.* — The author sees no reason for change in the opinion formerly expressed (Bigelow, 1904, pp. 365, 386) that echinoderms deserve little emphasis in an elementary course of biology or zoölogy. The school should have museum specimens, in formalin or alcohol, of American representatives of each of the five classes. The prices for good single specimens are reasonable. They may be purchased from the biological laboratories at Woods Hole, Mass., South Harpswell, Me., and other dealers (see Appendix III).

References for students: Kellogg, 1901, pp. 114–132; Jordan and Heath, 1902, pp. 140–150; Linville and Kelly, 1906, pp. 236–251; Colton, 1903, pp. 331–347; Jordan, Kellogg and Heath, 1903, pp. 150–160; Davenport, 1900, pp. 192–204; Davenport, 1911, pp. 246–259.

CHAPTER XIV

309. *Peripatus.* — The class Onychophora (genus Peripatus) is omitted from the text of the *Applied Biology* because of interest chiefly in morphological and phylogenetic interpretations that are beyond beginners in zoölogy.

310. *Crayfish.* — The value of this animal for elementary laboratory study is discussed by Bigelow, 1904, p. 358; an outline study, pp. 372–378; concerning specimens for study, pp. 404–406.

Dealers, 1910: The author deals with Burns, Chicago, and Knoll, New York (see list in Appendix III). Probably dealers in markets of most large cities can supply crayfishes in the autumn.

Homology. — The principle of serial homology as illustrated by crayfish appendages may be briefly presented, but probably means little to most beginners. Certainly, comparison of the first five pairs of appendages is useless in high schools.

Gill-currents are best shown by placing a crayfish on its back in one inch of water. Hold firmly but gently until quiet ("pseudo-hypnotism"), then release pressure and drop with a pipette some "carmine-water" near the last legs. Other finely divided insoluble pigments such as lampblack, India ink, gamboge mixed with water may be used.

References for students: Davenport, 1900, pp. 97–122; Davenport, 1911, Chapter IX; Kellogg, 1901, pp. 18–27,

153–154, 159–162; Jordan, Kellogg and Heath, 1903, pp. 116–118; Linville and Kelly, 1906, pp. 125–147.

References for teachers: Andrews, E. A., Article on "Keeping and Raising Crayfishes," in *Nature-Study Review*, Dec., 1906, pp. 296–301; also on "Crayfish Industry" in *Science*, June 29, 1906. Other references are given by Bigelow, 1904, p. 375.

315. References for students: Davenport, 1900, pp. 111–118; Davenport, 1911, pp. 135–147; Linville and Kelly, 1906, Chapter XIII.

316. References for students: Davenport, 1900, pp. 125–130; Davenport, 1911, pp. 153–160; Linville and Kelly, 1906, pp. 147–156.

318. *Spiders* may be preserved in ethyl alcohol, wood alcohol, or 2 or 5 per cent formalin solution. Since no dissection is required, specimens may be used for several years.

Demonstrate arrangement of the simple eyes of a spider under microscope. Emphasize comparison between spider and crayfish with regard to divisions of the body, simple and compound eyes, number of legs, legs with several joints, organs for breathing, external skeleton, appendages on abdomen. A comparative table should be begun with three columns, one to be filled out later after grasshopper is studied.

Habits.—Supplementary nature-study points on spiders. The secretion of spinning glands of some species forms webs for catching insects; in trap-door spiders it makes a silken lining to tunnels in soil and the framework of the door; in other species, it forms cocoons for eggs; in others the threads serve for uniting sticks, leaves, etc., to make a nest; in still others, threads are used for locomotion and may even form fluffy silken masses by which the spiders may float. Webs are of several types; see Davenport's *Zoölogy*, or Emerton's *Spiders*.

Poisonous effect of spider bites on men and large animals is never serious. Bites of ordinary spiders are no more dangerous than those of bees and mosquitoes.

References for students: Colton, 1903, pp. 54–60; Davenport, 1900, pp. 80–95; Davenport, 1911, pp. 102–119; Kellogg, 1901, 237–246; Jordan and Heath, 1902, pp. 133–139; Linville and Kelly, 1906, pp. 116–124.

References for teachers: Hodge, 1902, pp. 419–422; Holtz, 1908, pp. 246–250; Cornell Leaflets, 1904, pp. 65–66, 171–183; Emerton's *Spiders, Their Structure and Habits* and his *Common Spiders of United States;* Comstock's *Handbook of Nature-Study* (1912). A great systematic book by Professor Comstock is in press for publication in 1912.

321. *Myriapoda.* — Centipedes and millipedes are quite different in structure and some authors divide the class Myriapoda. Of course, such details are absurd for elementary work.

323. *Grasshopper.* — Materials: Bigelow, 1904, p. 407. Living grasshoppers may be kept in wire cages. Get a piece of wire-screen about 12×30 inches, roll so as to form a cylinder about 10 inches in diameter by 12 high, and cap one end with a disc of wire screen, cut the same diameter as the cylinder. Fine copper or soft iron wire is excellent for fastening the cap and side of the cylinder by sewing "over-cast" stitches through the meshes of the screen. Such wire cylinders may be inverted over potted plants, dishes with fresh grassy sod, and other objects suitable for insects.

Good methods on insect materials in general are given by Hodge, 1902, pp. 45–61; and Comstock's *Nature-Study,* 1912.

324. *Lepidoptera.* — Materials: Butterflies, tomato-worms and other larvæ; collected in midsummer and preserved in alcohol or formalin. Cocoons of large moths (cecropia, cynthia, etc.). They may be preserved in alcohol or formalin. Excellent practical notes are given by Dickerson, 1901, pp. 291–332. Some cocoons with living pupæ should be hung inside a cage described in § 323, and the emerging moths will be prevented from flying and breaking their wings. In such cages cecropia and cynthia moths pair naturally and produce fertile eggs, which may be kept in a box for some days until the larvæ hatch. Leaves of many trees and shrubs may serve as food; but no change of diet should be made after the larvæ begin on one kind of leaves. The author has seen cecropia caterpillars fed on leaves from privet hedges and reared to perfect pupæ. See Dickerson's book.

325. *Cicada.* — In the Southern States the life-history is only thirteen years long. A full and illustrated account of the Cicada is in Bulletin 14 of the U. S. Bureau of Entomology (for sale by Supt. of Documents, Washington, D.C.). There are also smaller pamphlets available free, especially annual circulars of the U. S. Dept. of Agriculture, giving the geographical location of expected broods.

326. *Insect classification.* — The grouping adopted in this section of the *Applied Biology* is easily learned and includes most of the economically important insects. The ten to fifteen additional orders adopted by various modern entomologists are chiefly of interest to taxonomists. The remarks in the paragraph on "netted-winged insects" in the *Applied Biology* will make it clear that insect classification which is more complicated than that of the important orders given in this section is undesirable for elementary students. The principle determining this simplifying of classification is stated by Bigelow, 1904, p. 281.

Wings of Hemiptera (p. 388 in *Applied Biology*). — The wings are absent in bed-bugs, cochineal-bugs, and scale-bugs. There are two wings in male coccids.

Number of insect species. — The figures on pp. 388 and 389 in the *Applied Biology* are very conservative estimates. The fact is that radical entomologists consider as species what conservative men call varieties. A radical species-splitter would no doubt make many species of dogs that are commonly considered varieties or breeds. However, even the lowest estimates of insect species are astounding.

327. *Useful insects.* — Reference: "Bees" (*Farmers' Bulletin 447*), Weed's *Farm Friends and Foes*, Part II (Heath, 1910). Smith's *Insect Enemies and Friends* (Lippincott, 1909).

328. *Injurious insects.* — Reference: See § 327 above.

329. *Mosquitoes.* — References: "Malaria" (*Farmers' Bulletin 450*), "Mosquitoes" (*Farmers' Bulletin 444*). See § 330 below.

330. *Flies.* — References: "House Flies" (*Farmers' Bulletin 459*); "How Insects Affect Health" (*Farmers' Bulletin 155*). "Typhoid Fly" (*Entomology Bulletin 78*); Doane's *Insects and Disease* (Holt, 1910); Howard's, *House Fly; Disease Carrier* (Stokes, 1911, $1.50).

328-330. *Economics of Insects.* — General references: Smith, J. B., *Insect Enemies and Friends* (Lippincott); Holtz, 1908, pp. 175-245; Hodge, 1902, pp. 62-89, 181-274; Davenport, 1900, pp. 66-73; Davenport, 1911, Chapter V; Cornell Nature-Study Leaflets; Comstock's *Handbook of Nature-Study*, 1912.

Many valuable pamphlets are issued by the United States Department of Agriculture; see list by Bigelow, 1904, p. 428; and consult circular on agricultural pamphlets classified for use of teachers. Also consult appendix of Holtz's *Nature-Study*, 1908.

CHAPTER XV

Mollusks

339. *Snails.* — The Helix pomatia is very common in France and Germany and is a pest in gardens and vineyards. Its scientific name means "apple snail." Purchased by the hundred, hibernating snails cost about $1.50 in New York from October to April. A few specimens by mail cost about four cents each. The author will gladly supply them at cost to teachers who need only a few for class work with the *Applied Biology.*

For methods of preparing specimens of snails, see Bigelow, 1904, p. 406; or *School Science*, January, 1903.

341. *Squid.* — It is more than useless, because inextricably confusing, to try to show beginners in zoölogy that the hind-end (fin-bearing) of the squid is dorsal and the head-end ventral. Simply use such terms as head-end and hind-end, upper and lower sides. For the morphological comparison with other mollusks, see Pratt's *Invertebrate Zoölogy*, 1901, p. 114, or zoölogical treatises.

For discussion of value of cephalopods in elementary zoölogy, see Bigelow, 1904, p. 367.

342. Roger's *Shell Book* is the only general book available; but experts say its text has many errors. However, the numerous illustrations make it useful.

CHAPTER XVI

VERTEBRATES

343. *Classes of vertebrates.* — The familiar division of vertebrates into five classes (fishes, amphibia, reptiles, birds, and mammals) appeals to the authors of the *Applied Biology* as best for an elementary course of zoölogy. It is used by A. Sedgwick in his well-known *Student's Text-Book of Zoölogy*, Vol. II, 1905. This is certainly best for beginners, who could not comprehend such a scheme as that which recognizes the fundamental distinction between Marsipobranchia or Cyclostoma (lampreys) and true fishes, thus making an additional vertebrate class. (See Parker and Haswell's *Text-Book of Zoölogy,* Vol. II, 1897.)

Likewise, it has seemed better in elementary zoölogy to follow those authors (see Sedgwick) who regard Vertebrata as a phylum and Chordata as a larger division including the vertebrate phylum and one or more protochordate phyla. This is simpler for beginners who will not soon study protochordates, and moreover will introduce no confusion if the protochordates are later studied in advanced college courses.

347. *Fishes.* — The sword-fish (p. 421 in *Applied Biology*) is a bony fish and has no close relationship to the saw-fish shark and saw-fish ray.

352. *Fossil reptiles.* — Page 427, second paragraph. Among long fossil reptiles that have been discovered, the Brontosaur in the American Museum of Natural History in New York is 66 feet long, the Diplodocus in the Pittsburgh Museum is over 80, and dinosaurs recently found in Africa are over 100 feet.

354-358. *Birds.* — Lantern slides for lectures on birds are sold by the National Association of Audubon Societies, 141 Broadway, New York City. Prices per slide, 30 cents plain and 70 cents colored. A list of more than two hundred slides will be sent on application. The same Association publishes excellent pictures of birds and leaflets for the use of teachers.

References: Probably the most useful book on structure and functions of birds is Beebe's *The Bird: Its Form and Function* (Holt, 1906, $3.50). See list of standard books on birds cited by Bigelow, 1904, p. 438. Also see Appendix of Holtz's *Nature Study.* Important books published since 1904 are: Hoffman's *Birds of New England and Eastern New York* (Houghton, Mifflin, $1.50); Trafton's *Methods of Attracting Birds* (Houghton, 1910, $1.20); Knowlton and Ridgway's *Birds of the World* (Holt, 1909, 873 pp., ill., $7). See also § 358 in this book.

356. "The alimentary canal is essentially the same in all birds." This refers to the general plan of structure, which has been variously modified in adaptation to food habits. (See Beebe, *The Bird.*)

358. *Economic Relations of Birds.* — This topic was omitted from the *Applied Biology* by an oversight in the last revision of the manuscript, and not because the authors regarded it as unimportant. Teachers should call the attention of students (1) to the enormous value of the domesticated birds for feathers, meat, and eggs; (2) to the usefulness of the insectivorous birds in helping to keep injurious insects in check; (3) to the hawks and owls that destroy rodents; and (4) to the æsthetic value of birds, because of their singing, beautiful plumage, and interesting habits. For this last reason alone most wild birds are well worth protection. The harmful habits of certain birds, notably the English sparrow, are well known. Weed and Dearborn's *Birds in Relation to Man* (Lippincott) is the most complete summary of the economic aspect of birds; but numerous pamphlets of the United States Department of Agriculture, Chapter I in Chapman's *Bird Life*, and chapters in many other books on birds discuss their economic relations. Other references: "Common Birds and Agriculture" (*Farmers' Bulletin 54*) *Educational Leaflets*, published by National Association of Audubon Societies, New York City; Weed's *Farm Friends and Foes*, Part IV (Heath); "Economic Value of Predaceous Birds" (*Yearbook Separate 474*, 1908).

359–362. *Mammals.* — Hornaday's *American Natural History* (Scribner's, $3.50), while dealing with all kinds of vertebrates, is especially good on mammals.

360. *One human species* (Homo sapiens) is mentioned on p. 437 in the *Applied Biology*. Some anthropologists classify some prehistoric skulls as belonging to another species (Osborn, *Age of Mammals*).

362. *Economic mammals.* — Reference: Weed's *Farm Friends and Foes*, Part V (Heath). The United States Department of Agriculture issues many pamphlets on farm mammals. Also, "Rabbits" (*Yearbook Reprint 452*, 1907); "Economic Value of Predaceous Birds and Mammals" (*Yearbook Reprint 474*); "Danger of Introducing Noxious Animals" (*Yearbook Reprint 132*, 1908); and others (see Circular 19, Div. of Pub., *Publications Classified for Teachers*. Standard books on mammals are cited by Bigelow, 1904, pp. 428, 439.

363–365. *Life-histories and sex-instruction.* — The text of the *Applied Biology* presents the life-histories of vertebrates more extensively than is done in other books for high schools, because the authors are convinced that biology is the only logical approach to the kind of sex-instruction that is now generally recognized as an important part of education.

If the *Applied Biology* is studied continuously from

Chapter I, then the students will have met the great facts of reproduction in parts of the following sections: 21, 22 (frog); 36 (sex-organs of frog); 56–63 (frog development); 27, 28 (plant); 74–81 (reproduction of bean plant); 196–211 (reproduction of seed-plants); 122 (comparison of animals and plants); 224–226 (fern); 232 (moss); 237–239 (simple plants); 244 (molds); 251 (yeast); 256 (bacteria); 267 (Paramecium); 270 (Amœba); 274 (malarial organism); 279 (Volvox); 281 (sponges); 283 (Hydra); 287, 288 (hydroids); 290 (Scyphomedusæ); 295 (tape-worm); 297 (Trichinæ); 299 (spontaneous generation); 301 (sand worm); 303 (earthworm); 309 (p. 363) and 310 (crayfish); 316 (barnacles); § 325 (insects); § 337 (clams); and finally 363–367 (vertebrates).

In high schools that are already taking advanced ground in that they are making their course of general biology apply to human life in the broadest possible way, there could be no more natural and scientific approach to the problems of human life-history than a review of the facts of reproduction in the order stated above. Perhaps it might seem to many teachers best to reverse the order. In that case, start a review with yeast, bacteria, and other simple plants; and then molds, moss, fern, bean, and other seed-plants. Likewise, review reproduction of one-celled animals, Hydra, "worms," crustaceans, insects, and other invertebrates; and finally take up §§ 363–367 on reproduction of vertebrates.

Such a review, culminating in the highest animals, will marshal the great facts of reproduction; and in the hands of a competent teacher will bring students to a serious and open-minded attitude which makes human life-history a natural culmination of biological study.

Just how much of the facts from human life should supplement such studies as §§ 363–367 is still a matter of discussion, but that considerable supplementing is now desirable is believed by the authors. The point of view quoted with approval by Bigelow, 1904, p. 285, is no longer tenable. Nothing is clearer than that it is inadequate to leave students to make their own application of biological ideas to human life. The demonstrated fact is that most of them do not, and so we need definite and applied teaching.

On a few points only do the authors of the *Applied Biology* feel confident as to the desirable work supplementary to biology in the line of sex-instruction:

(1) Co-educational college classes have often carried embryological study as far as outlined by § 367 in the *Applied Biology;* but for co-educational high schools it seems best that only teachers who have the rather rare ability to make their students feel that all biological facts are very serious, should attempt more than assigned reading of § 367. If a review of § 367 is made in class-work, it is better in the form of a straightforward lecture and not as an oral recitation. Of course no sudden change of method should be made for §§ 363–367 that will attract the attention of the students; it is certainly wiser to modify methods earlier in the course. One way of doing this is to assign sections for reading from time to time, and seemingly not to find time for recitations on them, or to lecture occasionally instead of holding an expected recitation.

(2) If the desirable instruction supplementary to § 367 and directly applied to human life is approved by the school authorities, separate classes for students of each sex are obviously necessary in either high school or general college. Here again in co-educational schools it is desirable that the attention of students should not be attracted by a radical change, such as a sudden division of a class, which would suggest to students that even biology dares

not frankly touch the problems of human life. It is better for the desirable division of a mixed class to be made on some reasonable pretext, such as the convenience of smaller groups for a promised review, which only the teacher knows is planned to lead up to human life-history. Tact and skill may enable the teacher to carry over to human life that attitude of frankness and open-mindedness which biology properly taught almost always develops in students.

Of course, the advanced student of human life must know that biological study soon comes face to face with some very difficult problems in connection with human life-history; but it is unfortunate for young people to have this impressed upon them before they have learned the deeper ethical and social reasons why, in general, there should be maintained a certain amount of reserve between the sexes in the consideration of some important problems of life. No educational theory or practice can possibly alter some fundamental aspects of human life; among them the psychical and physical relations of the sexes which nature has fixed immutably. Therefore, in presenting applied biological study as a part of general education, teachers do well to pause at the threshold of human life-history and recognize that there are good and sufficient reasons why very many important biological facts relating to reproduction should be frankly presented only to young people grouped in classes limited to one sex. This proposition seems as scientifically certain as the rotation of the earth.

(3) The third point that seems to the authors to be an established guide-post on the still incomplete pathway of sex-education is that the prominent biological facts of anatomy, physiology, embryology, and hygiene applied to human reproduction should be taught as supplementary to § 367, but *only by excellent teachers.* Just what this

means in detail is a problem on which the American Federation for Sex Hygiene will probably make some authori
tative announcement within a year; and the authors of
the *Applied Biology* hope to complete for publication a
manuscript that will suggest their views in detail as to
how § 367 should be expanded and supplemented.

(4) The fourth point on which the authors of the *Applied
Biology* are convinced is that sex-education will be inadequate until we know how to extend it safely and scientifically beyond the field of biology into ethics, sociology,
psychology, and æsthetics. However, this fact does not
interfere with the work of the teacher of biology, for
there is no question that the study of life-histories of
animals and plants is the only way for giving students
that outlook and attitude of mind that is absolutely
necessary for proper understanding of some of the most
complicated problems of human life; and it is the only
study that gives the fundamental knowledge of structure
and function of reproductive organs, which must be
understood before the individual can appreciate the problems of sex-instruction with reference to human life.
Hence, far from being discouraged because biology alone
is inadequate, the teacher, on the contrary, should find
stimulation in the fact that biology is the only scientific
approach to the greatest questions involved in human
life, and therefore the teacher of the science is working
on the foundations on which, before many years pass, we
may know how to build a satisfactory scheme of sex-
education. The most practicable step now possible in
the world-wide movement for sex-education is the development of the full possibilities of the biological studies that
touch the problems of reproduction. (Abstract of article
by M. A. Bigelow, in *Journal of Social Diseases*, October,
1912.)

References Relating to Sex-instruction

NOTE. — This selected list contains the books and articles most used by advanced students working with the author. A comprehensive work on sex-education now being prepared under the advice of a committee of the American *F*ederation for Sex Hygiene (New York, 105 W. 40th St.) will contain a more extensive bibliography.

American Society of Sanitary and Moral Prophylaxis (105 West 40th St., New York City) publishes six Educational Pamphlets :

No. 1 : *The Young Man's Problem.* Pp. 32. 10 cents.

No. 2 : *Instruction in the Physiology and Hygiene of Sex.* Pp. 24. 10 cents.

No. 3 : *The Relations of Social Diseases with Marriage and Their Prophylaxis.* Pp. 72. 25 cents.

No. 4 : *The Boy Problem.* Pp. 32. 10 cents.

No. 5 : *How My Uncle, the Doctor, Instructed Me in Matters of Sex.* Pp. 32. 10 cents.

No. 6 : *Health and Hygiene of Sex.* Pp. 32. 10 cents.

Journal of Social Diseases. Quarterly, $1.00 per year.

BIGELOW, M. A., "Relation of Biology to Sex-instruction in Schools and Colleges," *Jour. of Social Diseases*, **II**, 4, October, 1911.

CABOT, R. C., *Consecration of the Affections.* (Often Misnamed Sex-Hygiene.) Proc. of Fifth Congress of Amer. School Hygiene Assoc., III, 1911, p. 114.

EDDY, WALTER H., "An Experiment in Teaching Sex-Hygiene." *Jour. of Educ. Psychology*, II, 8, October, 1911, p. 440.

ELIOT, C. W., *School Instruction in Sex Hygiene.* Proc. of Fifth Congress of Amer. School Hygiene Assoc., New York, 1911, pp. 22–26.

GALBRAITH, ANNA, *Four Epochs of a Woman's Life.* Saunders, $1.50. (For mature women, especially mothers.)

HALL, G. S., *Adolescence.* Appleton, 2 vols., I, Ap. 469, 507–512. $7.50. (For parents and teachers.)

HALL, G. S., "Needs and Methods of Educating Young People in Hygiene of Sex." *Pedagogical Seminary*, 15 : 82–91, March, 1908.

HALL, G. S., "Teaching of Sex in Schools and Colleges." *Jour. of Social Diseases*, II, 4, October, 1911.

HALL, JEANNETTE, W., *Life's Story:* A Book on a Biological Basis for Girls of 10 to 17. B. S. Steadwell, La Crosse, Wis.; $.25.

HALL, W. S., *Developing into Manhood:* A Brief Book on a Biological Basis for Boys of 15 to 18. Association Press, New York City, $.25.

HALL, W. S., *From Youth into Manhood.* For Young Men and Older Boys. Association Press, New York City. $.50.

HALL, W. S., *Instead of Wild Oats: A Brief Book for Young Men on a Biological and Sociological Basis.* Revell & Co.

HALL, W. S., *Reproduction and Sexual Hygiene.* Wynnewood Pub. Co., Chicago. Pp. 150. $1.00. (For young men and fathers.)

HALL, W. S., "Social Hygiene: Its Pedagogic Aspects and Its Relation to General Hygiene and Public Health." *Nature-Study Review*, VI, pp. 33–39, February, 1910.

HALL, W. S., "The Teaching of Sexual Hygiene: Matter and Methods." *School Science and Mathematics*, X, pp. 469–474, June, 1910.

HENDERSON, C. R., *Education with Reference to Sex.* I, pp. 75, $.78 postpaid; II, pp. 100, $.80 postpaid. University of Chicago Press.

LOWRY, E. B., *Truths: Talks with a Boy* ($.50); *Confidences: Talks with a Young Girl* ($.50); *Herself: Talks with Women* ($1.10). Forbes, Chicago.

MARTIN, H. N., *Human Body, Advanced Course*, Holt, 1910. (Last chap. in any edition, but especially in Ninth, 1910, is excellent general account of anatomy, physiology, and hygiene of reproduction.)

MORLEY, MARGARET W., "Teaching Renewal of Life in Nature-Study." *Nature-Study Review*, November, 1907.

MORROW, PRINCE A., "Teaching of Sex-Hygiene." *Good Housekeeping*, March, 1912. Also pamphlets of the Society for Prophylaxis, see above.

MORROW, PRINCE, *Social Diseases and Marriage*, Lea Bros. 1909.

PARKINSON, W. D., "Sex and Education." *Educational Review*, 41, January, 1911; pp. 42–59. (Stands for ethical and æsthetic teaching.)

SALEEBY, C. W., *Parenthood and Race Culture.* Moffat, Yard & Co. 1909. (Popular account of eugenics.)

WILLSON, R. N., *The American Boy and the Social Evil.* Winston Co., 1909. $1.00. "Nobility of Boyhood," extract, $.50. For high-school boys.

Wood-Allen, May. *What a Young Woman Should Know.*
Vir Pub. Co. $1.00.
Also see § 491 in this *Manual* for books on heredity and eugenics.

367. It would be interesting to add to the legend of *F*ig. 154 that the embryo and all surrounding membranes shown developed from the fertilized egg-cell. There are thousands of cells in the stage represented.

369. Regarding the species of Homo, see note in § 360 in this *Manual.*

CHAPTER XVII

Human Anatomy and Physiology

370–490. *Order of study.* — This chapter on human structure and life-activities and the following chapter, which is largely hygienic, are preferably studied after Parts I, II, and III; but may be studied after Chapters I, II, III, and IV in order to make a short course that centers in human physiology. Still another possible arrangement for a very short course in the study of the human body is to start with §§ 370–373, then turn to §§ 38–41 and study tissues and cells, then §§ 42–55, which apply to the human body as well as to the frog. Substitute "human body" wherever the word "frog" occurs in §§ 42–55 (except in paragraph on heat on p. 52 and in references to respiration as done in part by frog's skin). Next study §§ 375–524. *F*inally study Chapter XVIII.

In either of the above plans for a brief course on human biology, the public hygiene part of Chapter XVIII ought to be introduced by study of bacteria (§§ 254–265).

Human and general biology. — On the educational relations of these two aspects of biology, see Bigelow, 1904, pp. 457–465. On relation of human aspect of biology to nature-study, see *Nature-Study Review*, Vol. II, pp. 67–72, February, 1906.

Emphasizing principles. — The authors of the *Applied Biology* are aware of much repetition in Chapter XVII of facts, especially physiological, stated in earlier chapters; but this has been deliberate and after due consideration leading to the conclusion that the great principles of biology must be repeated again and again from different points of view before they will become an integral part of the life and thought of the student. We cannot too often emphasize the statement that teachers of biology should aim to familiarize the student with the great ideas of the science. Certainly, familiarity does not often come from one momentary contact with things worth knowing in any field of knowledge. A further justification for repetition in this chapter is that here we are concerned with applying the great principles of biology to the human species.

Books Supplementary to this Chapter

For students: Eddy's *General Physiology;* Peabody's *Studies of Physiology;* Martin's *Human Body, Briefer Course;* and other standard high-school text-books of physiology.

For teachers: Hough and Sedgwick's *Human Mechanism;* Huxley and Lee's *Lessons in Physiology;* Martin's *Human Body, Advanced Course* (1910 edition); Howell's *Text-book of Physiology;* and Halliburton's *Handbook of Physiology.* See also appendix to this *Manual.*

370. *Anterior, posterior, dorsal, and ventral* should be applied as in the case of the frog (p. 26, *Applied Biology*). Anterior for ventral and posterior for dorsal in the older non-comparative human anatomy should not be used to confuse young students. The words anterior and posterior should be erased in description of *F*ig. 162, p. 512. More advanced students must also learn the older usage, which is still found in many text-books.

Selection of subject-matter. — The selection of facts for Chapters 17 and 18 in the *Applied Biology* is largely in harmony with the principles stated by Bigelow, 1904, pp. 465–472. The authors would advise schools to supplement many points, especially by experiments gleaned from standard text-books of elementary physiology and hygiene.

Teaching hygiene. We are now in a period of such rapid improvement of hygienic teaching, especially in the advanced nature-study and introduction to science in grammar grades and first year of high schools, that soon the time will come when only an outline of the biological relations of human life and other life will be needed in a course of applied biology. Hence it has seemed best to the authors to anticipate somewhat the culmination of this movement towards efficient hygiene teaching, and so they have omitted from the *Applied Biology* numerous elementary facts that belong in school years earlier than those for which that book is intended.

373. *Human structure.* — It is desirable to supplement the text by blackboard diagrams showing that the diaphragm divides the body-cavity into two, (1) the thoracic cavity with heart and lungs and (2) the abdominal cavity with the stomach, intestines, liver, kidneys, etc. Emphasize the fact that there are in the human body the same organs as in the frog, in the same relative positions, but that detailed study would show the human organs to be more complicated in structure. Emphasize the remarkable similarity of the general human structure to that of the frog, and indeed of all vertebrates.

374. *Cells.* — Emphasize the statement that life-activities are in living cells in all parts of the body and not simply in the brain, heart, and lungs, as was once supposed. Injury to any of these three organs may soon result in death, because of interference with the functions that serve the cells (see *Applied Biology*, § 436).

375. *Foods.* — It is not to be expected that students will do more than make a list of common animal and plant foods. Grouping according to constituent food-principles is included in §§ 378–381.

376. *Nutrients.* — Instead of this term some teachers prefer the term "food-compounds."

377. It seems best to introduce the chemistry of foods at this point instead of in the introductory chemical work in Chapter II; but on this matter authors disagree. See recent high-school books by Hunter (A. B. Co.), Peabody and Hunt (Macmillan), Payne (A. B. Co.), Sharpe (A. B. Co.).

377. *Fehling's reagent* for certain sugars (lactose, maltose, glucoses) may be purchased from chemists and kept in two bottles ready for mixing when needed. Solution A: copper sulphate, 35 grams in 200 cc. of water. Solution B: 48 grams of caustic soda (NaOH) and 173 grams of Rochelle salt (sodium potassium tartrate) in 480 cc. of water. When needed, mix $\frac{1}{2}$ a test-tube of B with $\frac{1}{5}$ a tube of A and $\frac{1}{3}$ a tube of water. Heat some of this mixture (blue in color) to boiling, and add the supposed sugar solution drop by drop. Red colored precipitate, especially after standing and cooling, indicates a reducing sugar, usually glucose in plant materials; but other tests are necessary to distinguish accurately the kind of sugar. Cane sugar (from beet, cane, or maple) does not give the red color.

For a simple test, make solution A as above and instead of B use a strong solution of caustic soda or potash. Mix and heat as directed for *F*ehling's test.

Eimer and Amend, New York, and doubtless other chemical supply companies, sell tablets for *F*ehling's solution. Price 20 cents per set of three vials of 25 tablets each.

Nitric acid test. — Cool test-tube before adding ammonia (p. 462 in *Applied Biology*).

378. *Carbohydrates.* — The formula for starch, dextrin and vegetable gums is $(C_6H_{10}O_5)_x$. Glycogen (animal starch) found in animal tissues (especially livers) should also be mentioned as a carbohydrate food. Cane-sugar is also called saccharose or sucrose.

379. *Digestible fats* (*e.g.*, in meat) and oils (*e.g.*, olive oil) are all liquid when they reach the intestine. Many old text-books called the fats "hydrocarbons"; but chemists now use this term for compounds of C and H; *e.g.*, benzine, naphtha, paraffin, and gasoline are mixtures of such compounds. The fats occurring in animals and plants are compounds of C, H, and O; *i.e.*, they are oxygen derivatives of hydrocarbons.

381. *Albuminoids* are so closely related to the typical albumins that they are now included among the proteins, under the name of sclero-proteins. Gelatin is digested like ordinary proteins. The one striking difference between albuminoids and the ordinary proteins is the absence of tyrosine and tryptophane, two amino-acids which enter into the composition of common proteins. Recent experiments have shown that by adding these to gelatin it may be made to take the place of ordinary protein food. However, for practical purposes gelatin, like fats and carbohydrates, must be used in mixed diet with the ordinary proteins obtainable from animal and plant tissues.

382. *Elements in human body.* — Examples of inorganic materials in the human body are calcium phosphate (abundant in bones and teeth), potassium and sodium phosphates in blood and tissues, calcium carbonate in bones and teeth, potassium and sodium carbonates and sulphates in blood and tissues, iron in hæmoglobin of the blood. The fourteen elements that appear to be essential in the composition of the human body are as follows, and the figures in parentheses indicate percentages: O (72), C (13.5), H (9.1), N (2.5), Ca (1.3), P (1.15); and the following are less than one per cent and important in the order stated: S, Na, Cl, *F*l, K, *F*e, Mg, Si. Note that over 97 per cent is composed of four elements, C, H, N, O.

388. *Epithelial cells* from the lining of the mouth may be obtained by gently scraping inside a cheek with a toothpick or flat piece of wood, and mounting the scrapings in a drop of water on an object-slide. Stain with iodine-eosin, or other anilin stains.

392. *Intestine.* — A detailed description of the small intestine seems unnecessary, but duodenum as the name of the part nearest the stomach might be introduced. Charts, manikins, illustrations in books, and blackboard diagrams will be useful in this and the two following sections.

396. *Digestive movements.* — Hough and Sedgwick's *Human Mechanism* contains diagrams illustrating muscular constrictions of the stomach and small intestine.

397. *Digestion of sugar.* — Only sugar of the grape sugar formula ($C_6H_{12}O_6$) is absorbed abundantly without digestion. Cane sugar is digested in the intestine from $C_{12}H_{22}O_{11}$ to $C_6H_{12}O_6$.

398. *Osmosis.* — For making apparatus for osmosis of digested foods, "gold-beater membranes for osmosis" are sold by Eimer and Amend, New York, at about $1 to $1.50 per dozen. Two bags, each about three inches long, may be made from each by carefully folding the open end, tying with coarse thread, and then cutting the bag in the middle. When water-soaked, the tied end will not leak. Always soak such membranes for ten minutes before using and test for leaking by wiping the outside dry and then filling with water. See also § 88 in this *Manual*.

A simple osmose-apparatus for use with this and similar experiments may be made as follows: Cut (with edge of a file) pieces of glass tubing about $\frac{3}{8}$ of an inch in diameter and three inches long. Heat one end of each piece until the glass is softened, and then press against a flat piece of metal so as to form a collar at the end of the tube. Now take a sheet of gold-beaters' membrane, parchment, or bladder, about two inches in diameter, fold it into bag form and tie around the collar of the glass tube, wrapping string firmly around several times and then tying.

The tube may be inserted through a hole in a piece of wood or cork which will form a convenient support over a small tumbler or bottle containing pure water into which the membrane should dip. Let the membrane soak in water fifteen minutes before using. Materials to be tested for osmosis may be poured into the tube by means of a rubber-bulbed pipette. A number of these pieces of apparatus should be made. If washed directly after using and then quickly dried, they may be used on many occasions without changing the membranes. Or after washing they may be preserved in a sealed Mason jar containing water with a teaspoonful (5 cc.) of commercial formalin in a quart of water. Probably a trace of boric acid, carbolic acid, or other antiseptic would preserve the membranes equally well.

416. *Energy.* — Probably the best elementary account of energy transformations in relation to food is given in Martin's *Human Body, Briefer Course*, in the chapter on "Why we Eat and Breathe." Also refer students to elementary text-books of physics. Clear ideas of energy conservation, storage, and transformation are essential to an understanding of even the elementary problems of human nutrition, and hence it is well worth while at this stage to digress into the field of physics far enough to present, or to review, the chief facts concerning energy that physiology needs.

418. *Protein oxidation.* — The availability of only 4 calories of the 5.6 in a gram of protein is due to the incomplete oxidation of that food in the living body. The nitrogenous excretions of cells are compounds of C H O N and may be completely oxidized in a chemist's bomb calorimeter. On the other hand, the fats and carbohydrates are largely oxidized to CO_2 and HO_2 in the living cells.

Oxidization at the low temperatures of living cells is probably due to enzyme action.

Energy value of food of farm animals is given in *Farmers' Bulletin 346*.

On chemists' methods of computing energy value of foods, see "Bomb Calorimetry" (*Bulletin 124, Bureau of Animal Industry*).

419. *Respiration calorimeter.* — An illustrated description of a calorimeter for human use is in *Yearbook Separate 539* (1910), free.

420. *Composition of typical foods* is given in the appendix of Sherman's *Chemistry of Food and Nutrition* (Macmillan, 1911, $1.50), an excellent handbook for students of dietetics and food economics, and also on the chemistry of digestion and metabolism. See the same book for comparative economy of foods, ranging from 8 cents per 3000 calories in wheat flour to $1.13 for eggs at 35 cents per dozen, $1.26 for lean beefsteak, and $1.90 for oysters.

Some interesting studies of foods and dietaries are in Sharpe's *Laboratory Manual*, Problem 42.

423. *Protein diet.* — A working man would require 720 grams of protein for necessary daily energy if he took no food but lean meat (middle of p. 500 in *Applied Biology*).

Growing children require more food than adults in proportion to weight. Some practical advice and references to books on feeding young children are included in Technical Education Bulletin No. 3 (10 cents), Teachers College, New York City.

. **424**. *Dietetics.* — On the problem of quantity of foods required, see Sherman's *Chemistry of Food and Nutrition*, New York, Macmillan, 1911, $1.50. A brief paper by the same author is in *Technical Education Bulletin No. 5* (10 cents), Teachers College, New York City.

425. *Respiration.* — See § 51.

429. *Air in lungs.* — About four to five per cent of the oxygen of the air is absorbed while in the lungs, and almost as much carbon dioxide is added

A strong adult at rest passes into and out of the lungs nearly 650,000 cubic inches of air in twenty-four hours (30 cu. in. $\times$ 15 times per minute $\times$ 60 minutes per hour $\times$ 24 hours = 648,000 cu. in.), while a man working hard will breathe over twice this amount.

To simplify the third paragraph in this section, erase "which contain . . . in the lungs." Also, 200, not 100 cubic inches *usually* remain in the lungs (last sentence); only 100 after *forced* expiration.

435. *Skin in excretion.* — The old idea that the skin is an essential excretory organ is largely based upon the fact that death once resulted from varnishing the entire skin and covering with gold-leaf. This coating was supposed to have prevented the escape of some very poisonous excretions. Later studies have shown that metallic coverings radiate heat rapidly and thus cause the lowering of the internal temperature far below the normal, resulting in "freezing to death." This does not mean 0° C. or 32° F., for a rabbit dies from low temperature if its internal heat is reduced below about 70° *F.* No human is known to have lived after reaching an internal temperature below 80° *F.* The skin might be completely covered with a varnish so as to close all pores of the sweat-glands and no harm result, provided that internal temperature remained normal. In short, the skin is a *heat-regulating* organ, not primarily an excretory organ, and there is no reason for supposing that it is necessary for eliminating any excretions that cannot be removed by the kidneys. It is clear, then, that excretion of water by the skin occurs only as a means for heat elimination.

447. *Internal heat.* — The importance of heat regulation may be further illustrated by reference to "sun-stroke" or "heat prostration," due chiefly or solely to high internal temperature. A man has remained eight minutes in dry air at 260° F., but the activity of the skin glands prevented internal overheating. An internal temperature above 104° *F.* means serious illness, **and** at about 108° a fatal result is almost inevitable.

CHAPTER XVIII

Biology Applied to Hygiene

448–462. *Personal hygiene.* — References: Pyle's *Personal Hygiene* (Saunders); Part II of Hough and Sedgwick's *Human Mechanism* (Ginn); Allen's *Civics and Health* (Ginn). All are adapted to advanced high-school students.

For young students: Ritchie's *Primer of Hygiene* (World Book Co.); Davison's *Health Lessons* (American Book Co.); Gulick's *Hygiene Series* (Ginn); Hutchinson's *Health Series* (Houghton, Mifflin).

463–481. *Stimulants and Narcotics.* — The laws of many States require high schools to give instruction concerning the physiological effects of alcohol, tobacco, tea, coffee, and drugs, usually in the first year. The topic is treated in the *Applied Biology* in accordance with laws in the States where the book may be used by teachers-in-training; and the briefer course will attempt to fit the laws relating to the first year of the high school. Obviously, more emphasis is given than would be demanded by its fair share of space in a book of applied biology. If either book is used under conditions that do not make such instruction mandatory, as in private schools, it will suffice to require a careful reading of the paragraphs 463–481 and to give a recitation including the chief points such as are summarized in § 478 and 479 of the *Applied Biology* and by Bigelow, 1904, p. 480. The same author gives (1904, pp. 472–485) a bibliography and summary of the relation of temperance instruction to high-school biology. There has been no notable change in the situation since 1904, except a widespread tendency to ignore the laws.

On the educational aspects of "temperance instruction," see Bigelow, 1904, pp. 472–485. Important additions to the literature list there given are: *F. L. Charles* in *Nature-Study Review*, Vol. IV, Dec., 1908, pp. 287–292. (Review of movement for change of law in Illinois.) "An official Letter," *Science*, Vol. XXVI, Oct. 4, 1907, pp. 443–444. (Instructions to publishers of text-books.)

Present Knowledge. Martin's *Human Body, Advanced Course*, Ninth (1910) edition, pp. 393–395. Howell's *Text-book of Physiology*, Third (1909) edition, pp. 887–889, have good summaries of the present state of physiological knowledge regarding alcohol.

465. *Effects of alcohol.* — While the first superficial effect of alcohol appears to be that of a stimulant, many authors regard it as really a narcotic, even in small doses, and believe that its apparent stimulating effect is due to inhibition or depression of certain control centers in the nervous system. For example, it is claimed by some physiologists that alcohol in small amounts causes rush of blood to the skin because it depresses the nerve center that normally controls vaso-constriction of cutaneous arteries; and likewise it increases the heart-beat because it depresses the center (cardio-inhibitory) that keeps the normal heart from too rapid action. It leads to wit and repartee because judgment and caution are inhibited. However, as a practical problem, small doses of alcohol result in certain stimulative effects, no matter how indirectly they may have been brought about; and so far as average intelligent readers can easily observe the effects they must be considered stimulating for small doses and narcotic for large ones. But this does not alter the fact that either a stimulant or a narcotic may be harmful — that is another problem.

467. *Alcohol a "poison."* — The authors of the *Applied Biology* are acquainted with various pharmacological literature on the classification of alcohol as a poison; but although they would gladly accept a scientific demonstration of the truth of that proposition, they are still unconvinced that there is anything to be gained in the line of temperance by publishing for lay readers the statement that alcohol is always a poison. Whatever may be the scientific conception of a poison, teachers of science have no moral right to overlook the popular understanding of the word; hence the attitude of § 467 in the *Applied Biology.*

481. *Drugs.* — Harmful effects of headache powders (*Chemistry Bulletin 126*, U. S. Dept. Agriculture, free; also *Farmers' Bulletin 377*). "Habit-Forming Agents" (*Farmers' Bulletin 393*).

482–490. *Public hygiene.* — Reference for students: Conn's *Bacteria, Yeasts, and Molds*, Chapters XV, XVI. Ritchie's *Primer of Sanitation* (World Book Co., Yonkers, N.Y.). See also §§ 448–462 in this *Manual.*

References for teachers: Part II of Hough and Sedgwick's *Human Mechanism;* Sedgwick's *Sanitary Science and Public Health;* Newman's *Bacteria and the Public Health*, Allen's *Civics and Health.*

484. *Preventing infection.* — References· "Typhoid" (*Farmers' Bulletin 487*); "House Flies" (*Farmers' Bulletin 459*). Also books mentioned above.

489. References· "Some Common Disinfectants" (*Farmers' Bulletin 345*). Chapter 17 in Conn's *Bacteria, Yeasts, and Molds in the Home.*

CHAPTER · XIX

Evolution and Heredity

491. For discussion of educational aspects of evolution, see Bigelow, 1904, pp. 286–289. For list of literature, see same author, pp. 429–432, and certain books named below. The following books published since 1904 are important :

Evolution

Campbell, D. H. *Plant Life and Evolution.* New York, Holt. 1911. Pp. 360. $1.60. (A popular historical-biographical account of biology and the evolution theory.)

Crampton, H. E. *Doctrine of Evolution.* New York, Columbia Univ. Press. 1911. Pp. 311. $1.50. (An interesting series of lectures in popular style.)

Geddes, P., and Thomson, J. A. *Evolution.* London, Williams. New York, Holt, 1911. Pp. 256. $.50. (One of the most readable general statements of evolution. Good . literature list.)

Locy, W. A. *Biology and Its Makers.* New York, Holt. 1908. Pp. 469, 123 figs. $2.75.

Metcalf, M. M. *Outline of the Theory of Organic Evolution.* New York, Macmillan. 1904. Pp. 204, 101 pls., 46 figs. $2.50.

Scott, D. H. *The Evolution of Plants.* London, Williams; New York, Holt. 1911. Pp. 256. Ill. $.50.

Thomson, J. A. *Darwinism and Human Life.* New York, Holt. London, Melrose. 1909. Pp. 245. $1.50. (Six introductory lectures on evolution. Bibliography excellent.)

Jordan, D. S., and Kellogg, V. L. *Evolution and Animal Life.* New York, Appleton. 1907. Pp. 489, 3 pls., 298 figs. $2.50.

Kellogg, V. L. *Darwinism To-day.* New York, Holt. 1907. Pp. 403. $2.00. (Excellent review of factors of evolution.)

Heredity

CASTLE, W. E. *Text-book of Heredity.* New York, Appleton. 1911. $1.50.

DAVENPORT, C. B. *Eugenics.* New York, Holt. 1910. $.50.

DAVENPORT, C. B. *Heredity in Relation to Eugenics.* New York, Holt. 1911. $2.00.

DONCASTER, L. *Heredity in the Light of Recent Research.* Cambridge University Press. New York, Putnams. 1911. Pp. 143. $.40.

KELLICOTT, W. E. *The Social Direction of Human Evolution.* New York, Appleton. 1911. Pp. 249. $1.50.

PUNNETT, R. C. *Mendelism.* London and New York. 1911. Pp. 192. $.50.

THOMSON, J. A. *Heredity.* London, Murray. New York, Putnam. 1907. Pp. 605, 49 figs. $3.50.

Also an excellent chapter in Geddes and Thomson's *Evolution.*

APPENDIX I

DRAWINGS AND NOTES

THE authors have attempted to designate in the *Applied Biology* only a few important drawings to be made by the pupils. Most teachers will probably require others. Since it is a question whether laboratory study of biology has not, in many high schools and colleges, become more of an art exercise than a scientific study, it seems best to require only drawings that represent structures of importance. The type of drawing best adapted to high-school biology is a simple line drawing without shading. Time spent in filling in details, *e.g.*, color spots of frog, is wasted so far as biology for general education is concerned, because in most cases such details are of little or no significance in relation to the great ideas of the science that it is the duty of the teacher and book to emphasize. Notes should be carefully made in all cases where record of laboratory work cannot be made in drawings. Clearly written and logical paragraphs should be required. On drawings and notes, see Bigelow, 1904, pp. 316–319; and Ganong, 1899, Chapter IV, or 1910, Chapter V, 89–108.

Students should study any illustrations in books which will help them understand structure. Some teachers will criticize this recommendation because they want the students to make original drawings and diagrams, a plan that is interesting in theory, but of very little significance in practice. If the teacher is chiefly interested in guarding against direct copying from books, then loose-leaf notebooks should be adopted and the sheets left on the teacher's desk at the close of each lesson. Of course, the average teacher is too busy to correct these sheets carefully, but in five minutes many sheets may be glanced over and marked with a rubber stamp so that substitution of later work is impossible. Do not pretend to correct the notes at the time of such brief examination, but simply aim to inspect and perhaps grade tentatively. In fact, I often wonder whether correction of mistakes in notes should not be done chiefly as a class exercise in which students either criticize each other's work or compare their own notes with some that have been selected and approved by the teachers. This would be an excellent experiment for a teacher with several sections of students taking the same course, for then results might be compared.   .

Notes should be concise, rarely in essay form, and as far as possible in outline and tabular form. There are occasions where copying of illustrations is desirable, provided that the student plainly indicates the source.

The figures in the *Applied Biology* were chiefly taken from standard biological books and have been credited, as far as possible, to the books used in copying. Teachers should refer to these figures for more complete descriptions than are given in text and legends in the text-book. The authors are of the opinion that, as a rule, the well-known figures from standard biological works are to be preferred to new ones that might be made to order. Moreover, good outline figures are usually far more significant, as illustrations correlated with the text, than photographs could be, although the latter might have an ornamental value.

APPENDIX II

Biological Books and Pamphlets

Teachers who wish suggestions regarding the selection of biological books are referred to the following authors: Ganong, 1910, Chapter VIII is an excellent account of botanical books. A less complete chapter is in the 1899 edition. Bigelow, 1904, Chapter X, has a selected list of books in zoölogy, physiology, bacteriology, general biology, and evolution. Many excellent books published since 1904 are named in this *Manual*. Lloyd, 1904, Chapter X, has a list of botanical books.

Methods of Teaching Biological Sciences

Ganong, W. G., *The Teaching Botanist*. Macmillan. 1899. Revised, 1910. $1.25. (Indispensable to teachers of botany and most of it useful for zoölogists.)

Lloyd, F. E., and Bigelow, M. A., *The Teaching of Biology in Secondary Schools*, Longmans, 1904, $1.50. Part I, *Teaching of Botany and Nature-Study*, by Lloyd; Part II, *Teaching of Zoölogy and Physiology*, by Bigelow.

Concerning Government Pamphlets

Many bulletins and other agricultural circulars mentioned in this *Manual* may be obtained free by writing to the United States Department of Agriculture, Washington, D.C., or to any congressman. Give numbers and titles of pamphlets wanted. Farmers' Bulletins are now sold at five cents each, but the Department and congressmen have free copies.

A bibliography classified for the use of teachers is published by the United States Department of Agriculture in Circular 19, Division of Publications.

A monthly list of agricultural publications is sent regularly to all applicants. It records both new and reprinted pamphlets.

Many of the more elaborate publications of biological value are for sale by the Superintendent of Documents, Washington, D.C. A list of such publications may be obtained from the Department of Agriculture.

There are many valuable pamphlets published by various colleges of agriculture and state experiment stations; *e.g.*, the famous Cornell Nature-Study Leaflets. Since in most cases these local bulletins cannot be widely distributed and the supply is limited, it has seemed best to give in this *Manual* references to those publications which are likely to be kept in print for the next five or ten years.

Books on General Biology for Teachers

NEEDHAM, J. G. *General Biology.* Ithaca, N.Y., Comstock Co., 1910. $2.00.

PARKER, T. J. *Elementary Biology.* Macmillan. 1897. $2.60.

SEDGWICK, W. T., and WILSON, E. B. *General Biology.* Holt. 1895. $1.75.

Books on Botany for Teachers

COULTER, J. M., BARNES, C. R., and COWLES, H. C. *Text-Book of Botany for Colleges.* 3 vols. A. B. Co. 1911.

CURTIS, C. C. *Nature and Development of Plants.* Holt. Revised, 1910. $2.50.

GANONG, W. G. *Laboratory Course in Plant Physiology.* Holt. $1.75.

Gray's New Manual of Botany. By Robinson and Fernald. A. B. Co. 7th ed., 1908. $2.50.

STRASBURGER and others. *Text-Book of Botany.* Macmillan. $5.00.

See also chapter on botanical books in Ganong's *Teaching Botanist*, 1911.

Books of Zoölogy for Teachers

See list given by Bigelow, 1904, pp. 419–421. The following are more recent zoölogies.

DREW, G. A. *Laboratory Manual of Invertebrate Zoölogy.* Saunders, 1907. $1.25.

HEGNER, R. W. *Introduction to Zoölogy.* Macmillan. 1910. $1.90.

OSBORN, H. *Economic Zoölogy.* Macmillan. 1908. $2.00.
SEDGWICK, A. *Students' Text-Book of Zoölogy.* 2 vols. Macmillan. Vol. I, $4.50; II, $5.00.
WEYSSE, A. W. *Synoptic Text-Book of Zoölogy.* Macmillan. 1904. $2.25.

Books of Physiology, Hygiene, and Bacteriology for Teachers

ALLEN, W. H. *Civics and Health.* Ginn. 1909. $1.25.
HALLIBURTON, W. D. *Handbook of Physiology.* (23d edition of Kirkes'), Blakiston. 1911. $3.00.
HOUGH, T. and SEDGWICK, W. T. *Human Mechanism.* (Part I, "Elements of Physiology," Part II, "Hygiene and Sanitation" are also published separately. $1.25 each.) Ginn. 1906. $2.00.
HOWELL, W. H. *Text-Book of Physiology.* Saunders. 1905. $4.00.
HUXLEY, T. H. *Lessons in Physiology.* Revised by F. S. Lee. Macmillan. 1900. $1.10.
JORDAN, E. O. *General Bacteriology.* Saunders. 1910. $3.00.
MARSHALL, C. E. *Microbiology.* Blakiston. 1911. $2.50.
MARTIN, H. N. *Human Body, Advanced Course.* Holt. Revised, 1910. $2.50.
PYLE, W. L. *Personal Hygiene.* Saunders. Revised, 1906. $1.50.

Arranging some books of physiology in order of difficulty for students, the list runs: Martin's *Human Body, Briefer Course* (for high school); Hough and Sedgwick or Huxley-Lee (for early college years); Martin's *Advanced Course* or Halliburton for a college course including histology, embryology, and physiology; and Howell for most advanced pure physiology.

Text-Books of Biology, Botany, Zoölogy, Physiology for High Schools

The following books should be in high-school libraries. Others are cited by Ganong, 1910, Lloyd, 1904, and Bigelow, 1904.

ANDREWS, E. F., *Practical Course in Botany.* A. B. Co. 1911. $1.25.
ATKINSON, G. F., *Botany.* Holt. 1910. $1.25.
ATKINSON, G. F., *First Studies of Plant Life.* Ginn. 1901. $.60.
BAILEY, L. H., *Beginners' Botany* ($.60), *Lessons with Plants* ($1.10), *Botany* ($1.10). Macmillan.

BERGEN, J. Y., *Foundations of Botany*, 1901 ($1.50); *Essentials of Botany* ($1.50); *Elements of Botany* ($1.30). Ginn.

BERGEN, J. Y., and CALDWELL, O. W. *Practical Botany.* Ginn. 1911. $1.50.

CLUTE, W. N. *Laboratory Botany.* Ginn. 1909. $.75.

COLTON, B. P., *Zoölogy.* (Text-book and practical book in one or two volumes.) Heath. 1903. $1.20.

COULTER, J. M., *Plant Relations* ($1.10); *Plant Structures* ($1.20); *Plant Studies* ($1.25); *Text-Book of Botany* ($1.25); Appleton.

DAVENPORT, C. B. and G. C. *Introduction to Zoölogy*, 1900. *Elements of Zoölogy*, 1911 ($1.20); Macmillan.

DAVISON, ALVIN, *Practical Zoölogy.* A. B. Co. 1906. $1.00.

EDDY, W. H., *General Physiology and Anatomy.* A. B. Co. $1.20.

HODGE, C. F., *Nature Study and Life.* Ginn. 1902. $1.50.

HOLTZ, F. L., *Nature Study.* Scribners. 1908. $1.50.

HUNTER, G. W., *Essentials of Biology.* A. B. Co. 1911. $1.25. (Revision of *Elements of Biology*.)

JORDAN, D. S., and Heath, H., *Animal Forms.* Appleton, 1902. $1.10.

JORDAN and KELLOGG, V. L., *Animal Life.* Appleton. 1900. $1.20. *Animal Life* and *Animal Forms* in one vol., *Animals*, $1.80. *Animal Studies.* $1.25.

KELLOGG, V. L., *Elementary Zoölogy.* Holt. 1901. $1.35.

LEAVITT, R. G., *Outlines of Botany.* A. B. Co. $1.00. (Really a revised Gray's *Lessons*.)

LINVILLE, H. R. and KELLY, H. A., *Text-Book of Zoölogy.* Ginn. 1906. $1.50. *Laboratory Guide*, $.35.

MARTIN, H. N., *Human Body, Briefer Course.* (Revised by Fitz). Holt. $1.25.

OSTERHOUT, W. J. V., *Experiments with Plants.* Macmillan. 1905. $1.25.

PEABODY, J. E., *Studies in Physiology.* Macmillan, $1.10.

PEABODY and HUNT, *Plant Biology.* Macmillan. 1912.

RITCHIE, J. W., *Primer of Hygiene* ($.40) *Primer of Sanitation* ($.50). (Intended for reading in grammar schools.) World Book Co., Yonkers.

SHARPE, R. W., *Laboratory Manual for Solution of Problems in Biology.* A. B. Co. 1911. $.75.

APPENDIX III

Biological Laboratory Equipment, Materials, and Methods

Laboratories and Equipment

References: Lloyd, 1904, Chapter IX, Ganong, 1910, Chapter VI; 1899, Chapter V, Bigelow, 1904, 409–416. See also sets of *School Science and Mathematics*, 1901 to present date; and *Journal Applied Microscopy*, 1898–1903.

Dealers in laboratory equipment are named by above authors, and they advertise in *Science, School Science and Mathematics,* and *Nature-Study Review.*

Dealers in Laboratory Equipment and Supplies

The author is best acquainted with the following firms who manufacture or deal in microscopes and other laboratory apparatus and supplies. The following abbreviations indicate the lines of stock prominently carried by each dealer : A, biological laboratory apparatus ; C, chemicals ; C P, apparatus for chemistry and physics ; F, laboratory furniture ; G, laboratory glassware ; M, microscopes, lenses, and accessories (such as slides and cover-glasses) ; O, optical apparatus (such as stereopticons, field-glasses). In writing for information, specify the lines you are interested in and ask for special catalogs.

Bausch & Lomb Optical Co., Rochester, N.Y. Branches in many cities. A, C, G, M, O, CP.

Spencer Lens Co., Buffalo, N.Y. A, G, M, O.

E. Leitz, 18th St., New York City. A, G, M, O.

Central Scientific Co., Chicago, Ill. A, C, CP, F, G, M, O.

Eimer & Amend, 18th St. and Third Ave., New York City. A, C, CP, G, M.

Cambridge Botanical Supply Co., Cambridge, Mass.

Kny-Scheerer Co., Ninth Ave., New York City. A, G, M, specialists in natural history specimens, skeletons, models, charts.

Williams, Brown & Earle, Chestnut St., Philadelphia. A, C, G, M, O.

Ward's Science Establishment, Rochester, N.Y. Prepared specimens of all kinds.

Whitall, Tatum & Co., Barclay St., New York City. G.

In many cities there are local dealers who sell at regular prices the goods of any manufacturing firms. Some local dealers known to the author are as follows : Philadelphia : A. H. Thomas Co., E. Pennock ; St. Louis : H. Heil Chemical Co. ; Pacific Coast : J. Caire, San Francisco, Braun-Knecht-Heimann Co., San Francisco, F. W. Braun Co., Los Angeles.

Biological Specimens for Study

Directions for collecting and names of some dealers are given by Ganong, 1899, Chapter VI ; 1910, VII (botanical); Lloyd, 1904, 222–228 (botanical); and Bigelow, 1904, Chapter IX (zoölogical).

Dealers

A. A. Sphung, N. Judson, Ind. (frogs).

F. J. Burns, W. South Water St., Chicago (frogs).

Louis Knoll & Sons, Washington Market, New York (crayfish, lobsters, clams, etc.).

Shawmut Market Co, State and Lake Sts., Chicago (crayfish in autumn).

Blackford's Market, Fulton Market, New York City (crayfish, crabs, clams, frogs).

Brimley Bros., Raleigh, N.C. (animal specimens).

Kny-Sheerer Co., New York City (living and preserved specimens for class work).

Western Biological Supply Co., Station A, Lincoln, Neb. (microscopic slides. Living and preserved biological materials).

F. Z. Lewis, Boys' High School, Brooklyn, N.Y. (microscopic preparations).

Marine Biological Laboratory, Woods Hole, Mass. (preserved marine organisms).

H. M. Stephens, Dickinson College, Carlisle, Penn. (preserved marine specimens).

Write to the professor of biology at the nearest college for names of local dealers.

Methods of Laboratory Work

REFERENCES: Ganong, 1899, Chapter III, or 1910, Chapter IV. Bigelow, 1904, Chapter III.

APPENDIX IV

A Year's Course in Biology

MANY teachers and principals have inquired concerning the place of the *Applied Biology* in high-school education, and this has suggested the possible value of a résumé of arguments for and against a year's course of biology in place of separate courses in botany and zoölogy.

The four-year curricula of most high schools is so arranged that there is great need of a one-year course of biological science adapted for the great majority of students. Here are the facts in support of this proposition : (*a*) It is generally admitted that four science courses, one for each year, offer the maximum amount of science desirable for the average secondary-school student. The fact is that the great majority of students in American schools get less than three science courses in four years of high school. (*b*) Chemistry, physics, botany, zoölogy, human physiology, earth science — a total of six — are the sciences which must be taken into consideration. Moreover, many schools have "general science" or "elementary science" in the first year. (*c*) There are two possible solutions, namely, election or concentration : Election means that students will fail to get a broad outlook on the general field of natural science, and possibly they may omit all biology. Concentration of the biological work into one course would leave biology, physics, chemistry, earth science — one course for each year of the high school ; or a choice of three of the four when "general science" occupies the first year. (*d*) A course in biology would greatly increase the number of students who have some knowledge of both the animal and plant aspects of life. With separate courses in botany and zoölogy, a large number of students who have time to study botany must omit zoölogy, or *vice versa*. This is very unfortunate, especially because both animals and plants are of such great interest as biology applied to human life. Moreover, the separate courses in botany and zoölogy do not properly develop the physiological and bacteriological phases in relation to human life.

The objection raised by some teachers who oppose a course in biology is that botany and zoölogy have developed into two sciences which are so distinct as to demand separate teaching. No doubt this statement is true in those all-too-numerous colleges where research for the few, rather than liberal biological training for the many, prevails; but high-school teachers should recognize that many college courses of technical botany and zoölogy are aimed at advanced courses and research and do not adequately present the great ideas or principles of the life sciences with reference to the needs of the average well-educated citizen. Both high-school and college teachers of biology need to study more seriously the problems of teaching an introductory biological course with reference to the needs of general education, rather than as a preparation for advanced technical courses. Viewed in this way, the teaching of introductory biological science in either school or college becomes the selection and presentation, not so much of the facts, as of the greatest ideas or principles which may be drawn from organized study of a series of plant and animal forms. These greatest ideas of biology are those which involve both animals and plants in their mutual relationships and in their bearing upon human life, and as separate sciences neither botany nor zoölogy can adequately present them. Herein is a decided advantage of a course of biology over any introductory courses of either botany or zoölogy. No botany or zoölogy course now in use as the introductory biological course in either high schools or colleges contains as many important facts and ideas for general education as are found in courses of general biology. Therefore the separate courses of botany or zoölogy are entirely inadequate for that vast majority of students who can take but one course and that the introductory, in biological science.

From numerous high schools and colleges has come the protest that botany and zoölogy (especially the special books) are so vastly rich in materials that even with a full year for each they cannot be "finished." The writer confesses that he has not been able to get into sympathy with this protest. Why should we want students to "finish" botany or zoölogy in one year, or even in five? We do not "finish" other subjects in high school or in college; but on the contrary, we select materials for well-rounded courses, some short and some long. Of course,

we cannot complete a wide survey of any of the biological
sciences in a single year, but there are great possibilities of
selection when our outlook on science for the average citizen
is that of liberal education as distinguished from technical
education. If one takes any current high-school or college
book on zoölogy or botany and goes through the pages critically
questioning each paragraph from the point of view of education
for general culture and information, he is amazed at the amount
of detailed matter which has no obvious relation to general
education. Eliminating such material of questionable value,
there is not left a superabundance of the essentials, the great
ideas, of the two phases of the science of ife needed for making
a single course which to the average educated citizen would be
more valuable than either botany or zoölogy studied without
reference to its sister science. And we must remember that
the sister sciences when presented in separate courses cannot
both be elected by numerous students in high schools and colleges.
Looking at biology from this point of view of liberal education
and laying entirely aside the practical problems of curricula,
which, as indicated above, are certainly tending to limit the
study of biology to a single year for the average student, a year's
course in biology is best because it forces selection of the valuable materials and develops most highly the important values
of biological science in general education.

The above discussion has touched only the informational
side of the values of biological study. Limitation of space
forbids appropriate discussion of the scientific discipline derivable
from the study; but on this point there is no argument against
the year in biology and in favor of botany and zoölogy taught
independently. On the contrary, we may expect to get more
valuable scientific discipline from the study of the more important
subject-matter which would be concentrated into a year of
biology. The only possible objections to this view are those
centered around the obsolete idea that science study must be
carried far into detail in order to give the best scientific discipline. This may be true from the research standpoint; but
as applied to the everyday life of the average cultured citizen
the results of study of biological details in schools and colleges
have been far from satisfactory. In all science teaching we seem
to be moving rapidly toward introductory courses which so
present information worth having that the discipline gained
will lead to greater application in practical life. Therefore,

so far as so-called scientific discipline is concerned, we may confidently assert that any possible advantage of one type of biological course over another will be in the one which offers the most valuable subject-matter, and this is, beyond serious doubt, a course which presents the general facts and ideas regarding living things — in short, a course of biology.

INDEX

The figures given below refer to pages, not sections, in this *Manual*. Teachers who refer directly from the *Applied Biology*, or from the table in the appendix of the *Introduction to Biology* should follow the section numbers rather than those of pages. Teachers who use this *Manual* in connection with other text-books should also examine the corresponding sections in the *Applied Biology*.

Made in the USA
Monee, IL
07 July 2026